my **revisi⏻n** notes

EDEXCEL GCSE (9–1)

GEOGRAPHY A

SECOND EDITION

Steph Warren

HODDER
EDUCATION
AN HACHETTE UK COMPANY

In order to ensure that this resource offers high-quality support for the associated Pearson qualification, it has been through a review process by the awarding body. This process confirms that this resource fully covers the teaching and learning content of the specification or part of a specification at which it is aimed. It also confirms that it demonstrates an appropriate balance between the development of subject skills, knowledge and understanding, in addition to preparation for assessment.

Endorsement does not cover any guidance on assessment activities or processes (e.g. practice questions or advice on how to answer assessment questions) included in the resource, nor does it prescribe any particular approach to the teaching or delivery of a related course.

While the publishers have made every attempt to ensure that advice on the qualification and its assessment is accurate, the official specification and associated assessment guidance materials are the only authoritative source of information and should always be referred to for definitive guidance.

Pearson examiners have not contributed to any sections in this resource relevant to examination papers for which they have responsibility.

Examiners will not use endorsed resources as a source of material for any assessment set by Pearson. Endorsement of a resource does not mean that the resource is required to achieve this Pearson qualification, nor does it mean that it is the only suitable material available to support the qualification, and any resource lists produced by the awarding body shall include this and other appropriate resources.

The Publishers would like to thank the following for permission to reproduce copyright material.

Photo credits p.8 *tl, bl & br* © Steph Warren; **p.16** *t* © Steph Warren; **p.25** © Steph Warren; **p.31** *b* © Steph Warren; **p.33** *t* © Steph Warren; **p.120** *t* © BEN STANSALL/AFP/Getty Images.

Acknowledgements
Every effort has been made to trace all copyright holders, but if any have been inadvertently overlooked, the Publishers will be pleased to make the necessary arrangements at the first opportunity.

Although every effort has been made to ensure that website addresses are correct at time of going to press, Hodder Education cannot be held responsible for the content of any website mentioned in this book. It is sometimes possible to find a relocated web page by typing in the address of the home page for a website in the URL window of your browser.

Hachette UK's policy is to use papers that are natural, renewable and recyclable products and made from wood grown in well-managed forests and other controlled sources. The logging and manufacturing processes are expected to conform to the environmental regulations of the country of origin.

Orders: please contact Hachette UK Distribution, Hely Hutchinson Centre, Milton Road, Didcot, Oxfordshire, OX11 7HH. Telephone: +44 (0)1235 827827. Email education@hachette.co.uk Lines are open from 9 a.m. to 5 p.m., Monday to Friday. You can also order through our website: www.hoddereducation.co.uk

ISBN: 978 1 4718 8725 3

Second edition © Steph Warren 2017

First published in 2013 by
Hodder Education,
An Hachette UK Company
Carmelite House
50 Victoria Embankment
London EC4Y 0DZ
www.hoddereducation.co.uk

Impression number   10 9 8 7

Year   2023

Cover photo © Oanh/Image Source/Corbis

Illustrations by Integra Software Services Pvt. Ltd., Pondicherry, India

Typeset in Bembo Std Regular 11/13 by Integra Software Services Pvt. Ltd., Pondicherry, India

Printed and bound by CPI Group (UK) Ltd, Croydon CR0 4YY

A catalogue record for this title is available from the British Library.

Get the most from this book

Everyone has to decide his or her own revision strategy, but it is essential to review your work, learn it and test your understanding. These Revision Notes will help you to do that in a planned way, topic by topic. Use this book as the cornerstone of your revision and don't hesitate to write in it – personalise your notes and check your progress by ticking off each section as you revise.

Tick to track your progress

Use the revision planner on page 4 to plan your revision, topic by topic. Tick each box when you have:
- revised and understood a topic
- tested yourself
- practised the exam questions and gone online to check your answers.

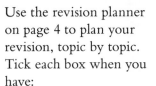

You can also keep track of your revision by ticking off each topic heading in the book. You may find it helpful to add your own notes as you work through each topic.

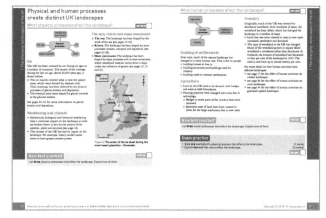

Features to help you succeed

Case studies and located examples

Revision notes on case studies and located examples are included so that you can give specific details in your answers.

Now test yourself

These short, knowledge-based questions provide the first step in testing your learning. Answers can be found online at: **www.hoddereducation.co.uk/myrevisionnotes**

Definitions and key terms

Here you will find some of the specialist geography terminology that you will need to know. The key terms are provided with clear, concise definitions.

Exam tips

Expert tips are given throughout the book to help you polish your exam technique in order to maximise your chances in the exam.

Exam practice

Practice exam questions are provided for each topic. Use them to consolidate your revision and practise your exam skills.

Online

Go online to check your answers to the exam questions at **www.hoddereducation.co.uk/myrevisionnotes**

My revision planner

REVISED | TESTED | EXAM READY

Now test yourself and exam practice answers at www.hoddereducation.co.uk/myrevisionnotes

Introduction

Revision technique

Revise actively

There is a large amount of factual detail that you have to remember for your Geography exam, and if you only sit and read through your work, you may not be able to remember it. Be active! Here are some activities that might help your memory:

Techniques for active revision

- Rewrite your notes on flash cards.
- Play music in the background. It helps your brain to focus and recall information.
- Try to teach a topic to another person.
- Talk to others – set up GCSE Geography discussion groups.
- Move around as you revise, wander around your house or garden.

Revision tips

- **Switch off the internet.** This stops you being distracted by social media sites and other websites.
- **Find your special place.** Allocate a room in your house as a working space. Your bedroom is probably not the best place!
- **Do short bursts of revision and reward yourself.** For example, do 20 minutes of revision for a reward of ten minutes on your favourite social media sites.
- Relate work to an **anagram** or draw a diagram to help you remember.
- Don't forget to **eat** plenty of fruit and **drink** plenty of water.
- Put information on **sticky notes** around your mirror. You will read them subconsciously as you clean your teeth.
- Use **practice papers** and **mark schemes** for revision, in order to become familiar with the wording of questions and how you answered or should have answered them.
- Think about where your weaknesses are and concentrate on revising for these topics.

Exam technique

Exam technique is all about how you complete the exam once you are in the exam room.

- You should always read the front of the paper. This can sometimes be done while you are waiting for the exam to start if the paper is facing upwards on your desk.
- When you are told to start the exam, do not waste time putting your name on the front. Instead, start immediately and then put your name on the paper at the end.
- Do not waste time looking around the exam room; keep yourself focused, concentrating at all times.

Components 1 and 2 exam techniques

Each of these papers has 94 marks and a time limit of 90 minutes. Therefore, you must try to keep to about one mark every minute.

Both of the papers have three sections. You should not have not been prepared for all of the questions. Look out for the options; choose the one you have been prepared for and put a cross in the box to indicate you have answered that question. Remember, the papers are marked electronically and a computer knows which question you have answered by the cross you put in the box.

Component 1 – answer question 1 in Section A and choose two from the other three questions. You should then answer all questions in Sections B and C.

Component 2 – answer all questions in sections A and B but in Section C answer the first question and then one from the other two questions.

In Section C, one of the questions, usually the last, has extra marks awarded to it for spelling, punctuation, grammar and specialist terminology. It is a good idea to do this section first or second, do not leave it until last; at the end of the exam, you may be rushing or tired, and your spelling and punctuation will not be as accurate.

Component 3 exam techniques

This paper has 64 marks and a time limit of 90 minutes. Therefore, you will have slightly more time for each mark, about one and a half minutes.

In Section A, you will choose from either rivers or coasts in the physical environments section and in Section B, you will choose from either urban or rural in the human landscapes section.

In Section C, all questions must be answered. The last question has extra marks awarded to it for spelling, punctuation, grammar and specialist terminology. It might be a good idea to do Section C first so that this question is not done in a rush at the end of the paper.

Command words

Here is a list of common command words which may be used on the exam paper. It is a good idea to underline the command words and any other key words in the question. The table is in order of the amount of marks which is likely to be awarded for questions starting with that command word. For example, the command word state would normally have one mark whereas evaluate would be eight or more marks.

> **Exam tip**
>
> Go through the sample assessment questions and any past papers on the Edexcel website to look at the range and balance of command words used.

Command word	Definition
Name, identify or state	These command words require you to answer briefly, usually just one word, and are usually only worth one mark. For example, 'give the grid reference for …' or 'name one type of sea defence'.
Define	You may be asked to define a term. The examiner will be expecting you to state the meaning of that term in a geographical framework.
Calculate	This asks you to produce an answer in numbers. It is a good idea to show your working out.
Draw or plot	These command words ask you to either do a diagram of a geographical feature or complete a graph.
Label	This is a simple descriptive comment which identifies something. If you are asked to annotate it means give a descriptive comment and an explanation.
Describe	This is a very common command word and requires you to give the main characteristics of something. Questions will often ask you to describe a photograph, a pattern on a graph or a map. You should write an accurate account of what you see but you do not need to provide a reason or justification for what you say.
Compare	This asks you to say in what way two or more things are alike, or different from each other. Remember to refer to both things and make a comment on their similarities and differences.
Explain	This is another very common command word. It is asking you to give reasons as to **why** something occurs.
Suggest/give reasons for	This is similar to 'explain' but sometimes there are varying reasons why something happens and there is not necessarily a right or wrong answer. The examiner will expect you to give more than one reason.
Examine	Look at the individual components and say why each one of them is important, how it contributes to the question and how the individual components work together and interrelate.
Assess	Decide the significance of something by studying all of the evidence. You should consider the importance of all the evidence and make sure you state which you think are the most important factors.
Discuss	If you are asked to discuss something, you will be expected to bring forward the important points of the argument which give the strengths and weaknesses of the different viewpoints.
Evaluate	Study the value of something. You should refer to all the factors which interrelate, draw on evidence and then make a judgement on which you think are the most important supported by evidence.

Mark schemes

Many questions are point marked. This means that each point you make is worth one mark. Longer questions which are worth more marks (eight marks normally) are marked using a levels mark scheme. Your answer is judged against the levels response and then a mark awarded in that level. It is a good idea to practise marking yours or your friend's responses using levels marks schemes so that you can improve your responses.

An example levels mark scheme for eight mark questions

Level/marks	Expected response
Level 1 1–3 marks	Attempts to answer the question with some understanding of the interactions and relationships between places, people and environments. The argument is incomplete and shows a lack of understanding. There is little evidence to support the answer. There is very little evidence for the judgements that are made.
Level 2 4–6 marks	Understanding is evident and there are logical connections between places, people and environments. There is still an imbalance in the argument but most of the understanding is relevant. Judgements are made that are occasionally supported by evidence.
Level 3 7–8 marks	There is understanding and logical connections between places, people and environment. The argument is well-developed and brings together relevant understanding supported by evidence.

An example levels mark scheme for SPGST

It is important to remember that some questions have spelling, punctuation, grammar and specialist terminology (SPGST) marks awarded to them. This means that the examiner will be looking carefully at your spelling, punctuation, grammar and the way that you use geographical terminology.

The questions that give marks for SPG are:
- Component 1 – the last part of the question in Section C
- Component 2 – the last part of the question in Section C
- Component 3 – the last question in Section C.

Marks	Commentary
0 marks	Nothing is written or the response does not relate to the question. There are errors in spelling, punctuation and grammar, which hinders the meaning of the response.
1 mark	There is reasonable accuracy in the spelling and punctuation. There is use of grammar so that meaning is not hindered. A limited range of specialist terms are used appropriately.
2–3 marks	There is considerable accuracy in the spelling and punctuation. There is use of grammar so that meaning is controlled. A good range of specialist terms are used appropriately.
4 marks	There is consistent accuracy in the spelling and punctuation. There is use of grammar so that meaning is effectively controlled. A wide range of specialist terms are used appropriately.

1 The changing landscapes of the UK

Geological variations within the UK

What are the characteristics and distribution of the UK's main rock types?

The UK's main rock types are:
- sedimentary (chalk and sandstone)
- igneous (basalt and granite)
- metamorphic (schists and slates).

> **Geology** is the science that deals with the history of the Earth, the rocks it is composed of and how it changes.
>
> **Fossil** is the remnant of an organism of a past time, such as a fish skeleton or a leaf imprint which has become embedded in a rock.
>
> **Crystal** is a material which is arranged in a regular form with definite lines of symmetry.
>
> **Texture** is the feel and appearance of a material.
>
> **Composition** is what a material is made up of.

Characteristics of metamorphic rocks
- They are formed from other rocks, either sedimentary or igneous.
- They are formed under great heat or pressure.
- Crystals can become arranged in layers such as slate which is formed from shale.
- For example, slate is extracted in Snowdonia.

Characteristics of igneous rocks
- They are formed from molten rock.
- They are made from randomly arranged crystals.
- They are very resistant rocks.
- They do not contain fossils.
- For example, granite at Haytor on Dartmoor.

Characteristics of sedimentary rocks
- Many have layers.
- They often contain fossils.
- They are composed of rounded grains pushed together.
- They vary greatly in colour.
- For example, chalk cliffs in Kent.

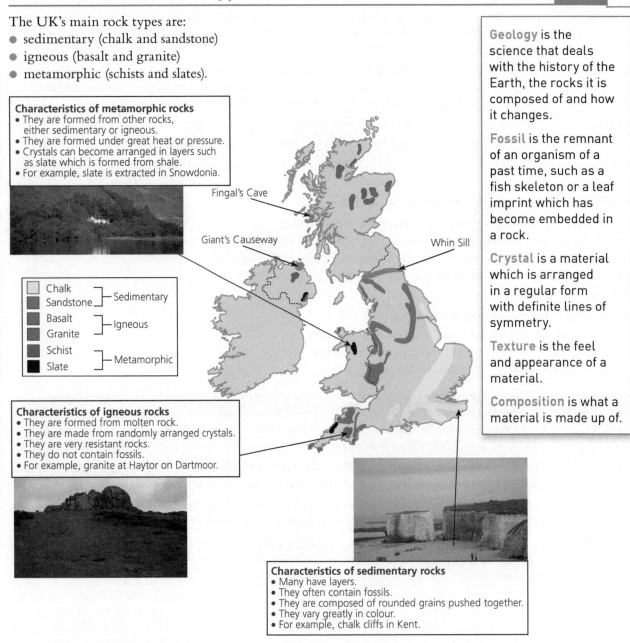

Fingal's Cave

Giant's Causeway

Whin Sill

Chalk / Sandstone — Sedimentary
Basalt / Granite — Igneous
Schist / Slate — Metamorphic

Figure 1 Simplified distribution and characteristics of the UK's main rock types

Now test yourself

TESTED

Name and locate an example of each of the UK's main rock types.

Exam tip

Ensure that you know examples of areas of the UK for all of the rock types mentioned.

Geological variations within the UK

What is the role of geology and past tectonic processes in the development of upland and lowland landscapes?

REVISED

Geology

Different types of rocks have varying resistance to physical processes.
- Igneous and metamorphic rocks tend to be more resistant than sedimentary rocks and therefore form highland areas.

- Sedimentary rocks are less resistant to physical processes and form lowland landscapes. They are much lower landscapes, though not necessarily flat, as they can contain rolling hills such as the North Downs.

N

Location on map	Name of river
A	River Tees
B	River Trent
C	River Thames
D	River Severn
E	River Exe

Highland areas
Rivers

Location on map	Name of upland
1	Northwest Highlands
2	Grampian Mountains
3	Pennines
4	Cambrian Mountains
5	Cotswolds
6	Chilterns
7	North Downs
8	South Downs
9	Exmoor
10	Dartmoor
11	Mourne Mountains

0   250 km

Figure 2 **UK upland areas**

Tectonic processes

Rocks which form the upland areas were made when the UK had tectonic activity.
- Igneous rocks were formed from the cooling of molten rock (magma).
- Metamorphic rocks were formed when sedimentary rocks were heated and compressed during tectonic activity.

Volcanic cones can still be seen in the UK landscape, for example Abbey Craig near Stirling is built on a volcanic plug.

Now test yourself

TESTED

1 Which rock type is likely to be found in lowland landscapes?
2 Why is this rock type likely to be found here?
3 Where in the UK can you find upland landscapes?
4 What role did tectonic activity have in forming the UK's landscape?

Physical and human processes create distinct UK landscapes

What physical processes affect the landscape?

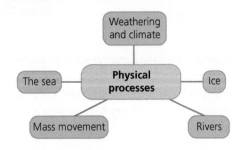

Ice

The UK has been covered by ice during ice ages on a number of occasions. The extent of the coverage during the last ice age, about 20,000 years ago, is shown below.

- The ice mainly covered what is now the upland areas, which were formed by resistant rock. Their landscape has been defined by the physical processes of glacial erosion and deposition.
- The lowland areas were shaped by glacial outwash as the glaciers melted.

See pages 30–34 for more information on glacial erosion and deposition.

Weathering and climate

- Mechanical, biological and chemical weathering have a continual impact on the landscape as rocks are broken down in situ by the actions of the weather, plants and animals (see page 12).
- The climate of the UK has had an impact on the landscape. For example, heavy rainfall causes rivers to have greater erosive power.

The sea, rivers and mass movement

- **The sea:** The landscape has been shaped by the work of the sea (see pages 13–16).
- **Rivers:** The landscape has been shaped by river processes: erosion, transport and deposition (see pages 21–25).
- **Mass movement:** The landscape has been shaped by slope processes such as mass movement, where weathered material moves down a slope under the influence of gravity (see pages 12, 21 and 31).

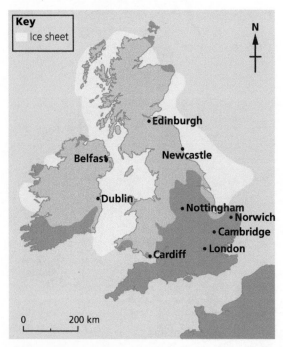

Figure 3 **The extent of the ice sheet during the most recent glaciation – Devensian.**

Now test yourself

List **three** physical processes that affect the landscape. Explain one of them.

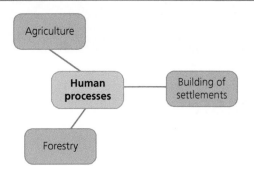

Building of settlements

Over time, much of the natural landscape has changed to a more human one. This is due to people

- building houses to live in
- building structures and buildings used for industry
- building roads to connect settlements.

Agriculture

- Land in the UK used to be farmed, with hedges and walls as field boundaries.
- Farming practices have changed over time due to technology.
  - Hedges in some parts of the country have been removed.
  - Extensive areas of land have been created to allow for the large machinery that is now used.

Forestry

Originally, much of the UK was covered by deciduous woodland. Over hundreds of years, the woodland has been felled, which has changed the landscape in a number of ways:

- Land that was once covered in trees is now open moorland, settlement and farmland.
- The type of woodland in the UK has changed. Much of the woodland grown to replace felled woodland is coniferous rather than deciduous. In Scotland, the amount of woodland had decreased to four per cent of the landscape by 1900. The total is now back up to almost twenty per cent.

For more detail on how human activities have affected landscapes:

- see page 17 for the effect of human activities on coastal landscapes
- see page 26 for the effect of human activities on river landscapes
- see page 34 for the effect of human activities on glaciated upland landscapes

Now test yourself

TESTED

List **three** human processes that affect the landscape. Explain one of them.

Exam practice

1 State **one** example of a physical process that affects the landscape. (1 mark)
2 Explain **one** way that rivers affect the landscape. (2 marks)

ONLINE

2 Coastal landscapes and processes

Physical processes shape coastal landscapes

What are the physical processes at work on the coast?

What are the main types of weathering?

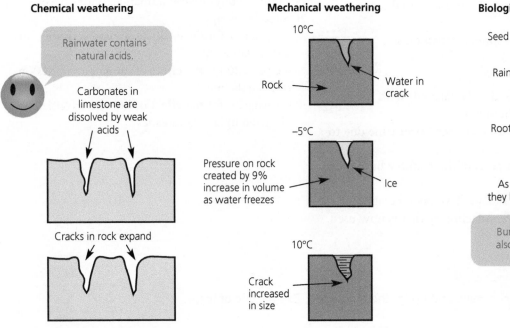

Chemical weathering

Rainwater contains natural acids.

Carbonates in limestone are dissolved by weak acids

Cracks in rock expand

Mechanical weathering

10°C

Rock

Water in crack

−5°C

Pressure on rock created by 9% increase in volume as water freezes

Ice

10°C

Crack increased in size

Biological weathering

Seed falls into crack
↓
Rain causes seedling to grow
↓
Roots force their way into cracks
↓
As the roots grow they break up the rock

Burrowing animals also break up rock.

Figure 1 **Types of weathering**

1 State **three** types of weathering.
2 Explain the process of mechanical weathering.

> **Weathering** is the way that rocks are broken down in situ.
>
> **Mass movement** is when material moves down a slope, pulled down by gravity.

What is mass movement?

Mass movement is when material moves down a slope due to the pull of gravity. You need to know about slumping and sliding.

Slumping and sliding
- A large area of land moves down a slope.
- It is common on clay cliffs.
- Dry weather makes the clay contract and crack.
- When it rains, water gets into the cracks.
- The soil becomes saturated.
- A large piece of rock is pulled down the cliff face.
- It has slipped on the slip plane of saturated rock.

Exam practice

Explain **one** type of mass movement. (3 marks)

What is erosion, transport and deposition?

The coast is under continual attack from waves at the base of the cliff and other processes on the cliff face such as weathering and mass movement.

Exam tip

Make sure you know all of the terms in the key terms box as you could be asked to define them in an exam.

Exam tip

Remember, attrition is not a type of erosion that erodes cliffs.

Erosion is when rocks are worn down by an agent of erosion such as water.

Hydraulic action occurs when the compression of air in cracks puts pressure on the rock and causes pieces of the rock to break off.

Abrasion occurs when sand and pebbles carried in waves are thrown against the cliff face.

Solution is when chemicals in sea water dissolve certain rock types, such as chalk.

Attrition is the breaking up of rocks and pebbles in the waves. The movement of waves means that rocks are continually knocked against each other, removing any sharp edges, to produce smooth pebbles and, eventually, sand.

Transportation is the movement of sand and pebbles by the sea.

Deposition is the putting down of sand and pebbles by the sea.

Traction is the rolling of large sediment such as pebbles along the sea bed.

Saltation occurs when small pieces of shingle or large grains of sand are bounced along the sea bed.

Suspension is when small particles such as sand and clays are carried in the water. This can make the water look cloudy especially during storms or when the sea has lots of energy.

Swash is the forward movement of a wave.

Backwash is the movement of the wave back down the beach.

Fetch is the distance over which the wind blows over open water.

The process of longshore drift

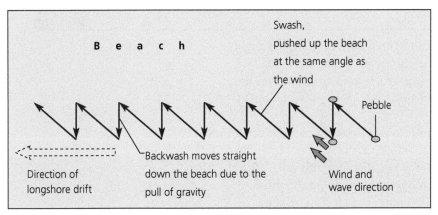

Figure 2 **The process of longshore drift**

How does geological structure have an impact on landforms?

Well-jointed rocks or ones with lines of weakness will erode more quickly as the waves exploit them.

Concordant coastlines have rocks that are parallel to the coastline. They have alternate layers of hard (more resistant) and soft (less resistant) rock. The hard rock acts as a barrier to the erosive power of the sea.

Impact of geological structure on landforms

Cliffs made from resistant rock, like granite, will erode more slowly than cliffs made from less resistant rock, such as clay.

Discordant coastlines have rocks which are at right angles to the sea. If there are alternate layers of hard and soft rock, the soft rock will erode more quickly forming bays with the hard rock forming headlands.

What types of waves are there?

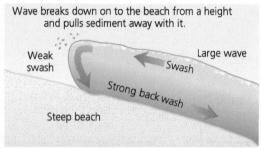

Wave breaks down on to the beach from a height and pulls sediment away with it.

Weak swash

Large wave

Swash

Strong backwash

Steep beach

Figure 3a **A destructive wave**

Wave breaks forwards on to the beach and so builds up sediment.

Strong swash

Small wave

Weak backwash

Gentle beach

Figure 3b **A constructive wave**

Now test yourself

State **two** differences between constructive and destructive waves.

How does the UK's weather and climate affect rates of coastal erosion?

- **Seasonality** – in the winter the differences between day and night time temperatures can cause freeze-thaw weathering on a cliff face.
- **Storms** – more storms are occurring which have an impact on the landforms of the coastline as storm waves are more powerful agents of erosion.
- **Prevailing wind** – this is the wind from the southwest. The coastlines of Cornwall and Devon experience winds that may have blown for several thousand kilometres across the sea. These winds have a long fetch. The longer the fetch, the stronger the wind and the more powerful the wave, thus increasing the rates of erosion on these coastlines.

Coastal erosion and deposition create distinctive landforms within the coastal landscape

What landforms are created by coastal erosion?

REVISED

Distinctive and dynamic landforms are formed by destructive waves. These include: cliffs and wave-cut platforms; headlands and bays; caves, arches and stacks.

Erosional features – cliffs and wave-cut platforms

Above the wave-cut notch an overhang develops. As the notch becomes larger the overhang will become unstable. This is because of its weight and the lack of support. In time the overhang will fall due to the pull of gravity.

In these boxes the explanation is underlined.

As the width of the platform increases the power of the sea decreases, because it has further to travel to reach the cliff and the water is shallower causing more friction.

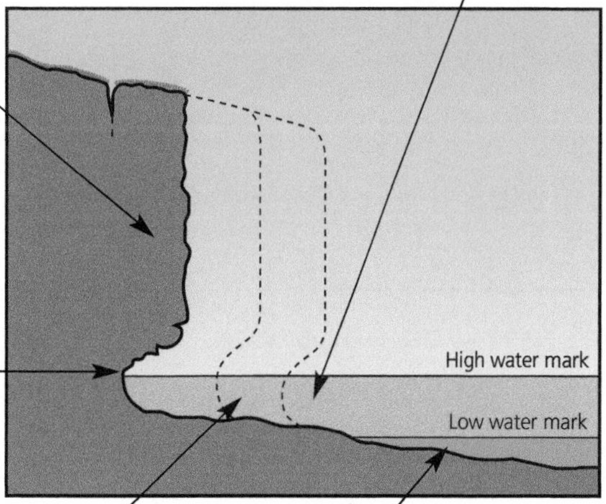

The cliff is eroded at the bottom by **corrasion. This is pebbles carried by the sea which are thrown against the cliff by the breaking wave, knocking off parts of the cliff.** In time a wave-cut notch is formed.

High water mark

Low water mark

The sea continues to attack the cliff in this way and the cliff retreats.

The remains of the cliff, now below the sea at high tide, form a rocky wave-cut platform. The platform will also contain the boulders which have fallen from the cliff.

Figure 4 **The formation of cliffs and wave-cut platforms**

Erosional features – headlands and bays

Headlands and bays form due to different rock types.
- They only occur on coastlines where soft and hard rocks are found at right angles to the sea.
- The soft rock erodes more quickly than the hard rock, forming bays.
- The hard rock is more resistant and sticks out as headlands.
- You should also refer to the processes of erosion.

Now test yourself

Explain the formation of headlands and bays.

TESTED

Exam tip

A top-grade candidate answering a question on headlands and bays may also refer to the fact that erosion eventually becomes greater on the headlands because the bays have retreated and the headlands are more exposed.

Erosional features – caves, arches and stacks

1 Sea exploits fault in a cliff face, using erosional processes such as hydraulic action.

2 In time the fault will widen to form a cave. If the fault is in a headland, caves are likely to form on both sides. When the backs of the caves meet, an arch is formed.

4 This leaves behind a column of rock not attached to the cliff, known as a stack, which is undercut at the bottom forming a wave-cut notch.

3 The sea will continue to erode the bottom of the arch using abrasion. It will collapse in time, as it is pulled down by the pressure of its own weight and gravity.

Figure 5 **The formation of caves, arches and stacks**

Exam practice

Examine how physical processes work together in the formation of the stack shown in Figure 5.

(8 marks)

ONLINE ☐

Exam tip

When answering questions about landform formation:
- always include an explanation of processes
- describe in detail and explain what happens in a sequence of events
- drawing a diagram will help your answer.

What landforms are created by coastal deposition?

REVISED ☐

Depositional features – beaches

Found on straight coastlines where longshore drift is happening

Found in bays where the sea is shallower so the waves lose their energy and deposit what they are carrying

Beaches

Made up of sand and pebbles

Formed by constructive waves

Depositional features – spits and bars

- Spits are narrow stretches of sand and pebbles that are joined to the land at one end.
- If you are asked to explain the formation of a bar you need to discuss spit formation and then add the following points:
  - Bars are spits which go across a bay.
  - This is only possible if there is shallow water and no river entering the sea.

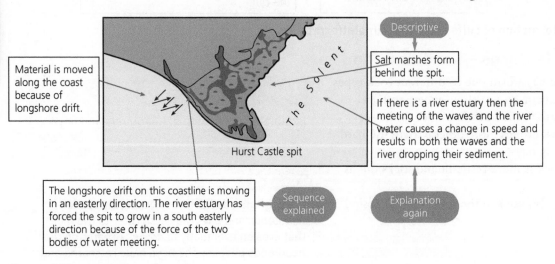

Material is moved along the coast because of longshore drift.

Descriptive

Salt marshes form behind the spit.

If there is a river estuary then the meeting of the waves and the river water causes a change in speed and results in both the waves and the river dropping their sediment.

Hurst Castle spit

The longshore drift on this coastline is moving in an easterly direction. The river estuary has forced the spit to grow in a south easterly direction because of the force of the two bodies of water meeting.

Sequence explained

Explanation again

Figure 6 **The formation of a spit**

Human activities can lead to changes in coastal landscapes which affect people and the environment

How have human activities affected coastal landscapes?

REVISED

- Urbanisation has had a visual impact on the coastal landscape. The building of settlements including industry and coastal defences has had an impact on the appearance of the coastal landscape. Urbanisation has also had an impact on wading birds and wildlife that have lost their homes in the coastal landscape. It has also had an impact on coastal processes such as longshore drift.
- Agriculture has had an impact because of the draining of coastal wetlands for farming activities.

What are the effects of coastal recession on people and the environment?

REVISED

England has 2,800 miles of coast of which 1,100 miles are classed as being at risk from erosion. Coastal recession affects both people and the environment.

- **Effects on people** – loss of homes. An example of this is Seaton in Devon which will not be defended after 2025 because it is a small town.

- **Effects on the environment** – loss of land. An example of this is the National Trust area in Dorset known as Golden Cap close to the village of Seatown. The cliff has receded 40 metres in the last 20 years.

> Coastal recession is the gradual movement backwards of the coastline, which is the dividing line between land and sea.

Happisburgh – a doomed village!

Since 1995, 25 properties and the village's lifeboat launching station have been lost to the sea. The village contains 18 listed buildings including a Grade 1 listed church which is estimated to be in the sea by 2020. The lives of the villagers are totally dominated by their struggle against the sea.

Train passengers get a shower

Passengers on the train travelling from Exeter to Plymouth and Penzance regularly get a shower as the sea washes over the tracks. On one occasion, 160 passengers were stranded in a train for four hours while the sea washed over them because the train's electrics were not working. The situation was made worse in February 2014 when 80 m of sea wall beneath the railway collapsed in a storm and the railway line was left with no land beneath it. The line was closed until early April.

Now test yourself

TESTED

Using examples, explain the effects of coastal recession.

What are the effects of coastal flooding on people and the environment?

In England 2.1 million properties are at risk from flooding. Nearly 50 per cent of these properties are at risk from flooding by the sea.

Coastal flooding is the inundation of land close to the sea by seawater.

Effects on people
- Damage to people's homes and belongings from water.
- Drowning.
- Contamination of fresh water supplies by sewage water.
- Bridges and roads can be washed away.
- Disruption to gas and electricity supplies.

Effects on the environment
- Loss of land to the sea.
- Crops ruined by sea water.
- Trees and vegetation washed away.
- Forming of new habitats due to inundation of flood water.

Sea surges in!

In December 2013, the combination of a high spring tide, an area of low pressure and strong northerly winds caused a storm surge that flooded much of the 45 miles of coastline of North Norfolk. People were warned and evacuated but 152 houses and businesses were damaged.

Pebbles hit the bar!

In February 2014, the Cove House Inn on the Isle of Portland was hit by 60 ft waves which crashed over the top of the building throwing sea water and pebbles into the bar.

Now test yourself

Using examples, explain the effects of coastal flooding.

What are the main types of hard and soft engineering used on the coastline of the UK?

A number of different types of coastal defences are used to defend the UK coastline; they can be hard or soft engineering. What are their advantages and disadvantages and how do they change coastal landscapes?

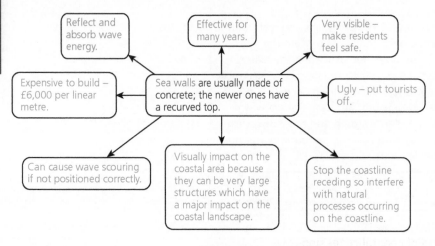

Soft engineering is a method of coastal management which works or attempts to work with the natural processes occurring on the coastline.

Hard engineering is a method of coastal management which involves major construction work, for example sea walls.

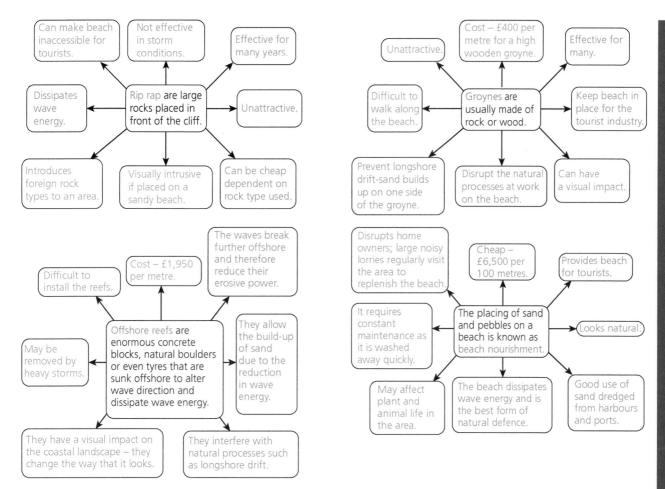

Can make beach inaccessible for tourists.

Not effective in storm conditions.

Effective for many years.

Dissipates wave energy.

Rip rap are large rocks placed in front of the cliff.

Unattractive.

Introduces foreign rock types to an area.

Visually intrusive if placed on a sandy beach.

Can be cheap dependent on rock type used.

Unattractive.

Cost – £400 per metre for a high wooden groyne.

Effective for many.

Difficult to walk along the beach.

Groynes are usually made of rock or wood.

Keep beach in place for the tourist industry.

Prevent longshore drift-sand builds up on one side of the groyne.

Disrupt the natural processes at work on the beach.

Can have a visual impact.

Difficult to install the reefs.

Cost – £1,950 per metre.

The waves break further offshore and therefore reduce their erosive power.

May be removed by heavy storms.

Offshore reefs are enormous concrete blocks, natural boulders or even tyres that are sunk offshore to alter wave direction and dissipate wave energy.

They allow the build-up of sand due to the reduction in wave energy.

They have a visual impact on the coastal landscape – they change the way that it looks.

They interfere with natural processes such as longshore drift.

Disrupts home owners; large noisy lorries regularly visit the area to replenish the beach.

Cheap – £6,500 per 100 metres.

Provides beach for tourists.

It requires constant maintenance as it is washed away quickly.

The placing of sand and pebbles on a beach is known as beach nourishment.

Looks natural.

May affect plant and animal life in the area.

The beach dissipates wave energy and is the best form of natural defence.

Good use of sand dredged from harbours and ports.

Now test yourself

TESTED ☐

What are the differences between hard and soft engineering techniques?

Exam practice

Explain **one** way that offshore reefs help to protect coastlines. (2 marks)

ONLINE ☐

Exam tip

Read the question carefully and don't get caught out. Underline the key words in the question.

Exam tip

Questions could be based on the advantages and disadvantages of one of these techniques or could be more general about the advantages and disadvantages of hard engineering and soft engineering techniques.

Interaction between physical and human processes leads to distinctive coastal landscapes

Located example: Isle of Purbeck, Dorset

Significance of the location

The Isle of Purbeck is part of the Jurassic Coast which is famous for fossils and distinctive coastal features such as headlands and bays. It is a concordant coastline to the south of the promontory and a discordant coastline to the east (see page 14).

Physical process	Human process
Coastal erosion – Ballard Down is constantly changing shape due to erosion and weathering. Originally, there were two stacks off the coast, called Old Harry and Old Harry's Wife. In 1896, the stack that was referred to as Old Harry's Wife collapsed forming a stump.	**Coastal defences** – in 2005 to 2006 new coastal defences were built in Swanage Bay consisting of 18 groynes and beach nourishment. This caused a change to the area as a new higher beach was created although it will have to be replenished every 20 years due to the erosion rates in the area.
Landslips – the coastline to the south of Ballard Down has frequent landslips causing the coastal path to have to be redirected on a number of occasions.	**Human development** – the building of Swanage town, especially the houses and hotels on the cliff, have made the problem of land slipping in the area worse.
	Tourism (Swanage and Studland Bay to the north of the area) – the beach at Studland is owned and managed by the National Trust. The area is protected from excessive tourist damage by limiting the parking available and therefore the number of people who can access the beach. The sand dunes are also protected by being fenced off. In this way, changes to the area as a result of human processes is being managed.

Figure 7 **Physical and human processes that have shaped the coastal landscape of the Isle of Purbeck**

Now test yourself

TESTED

Explain how physical processes formed the discordant coastline of the Isle of Purbeck.

Exam practice

For a named example, examine how the distinctive coastal landscape is the outcome of physical and human processes.

(8 marks)

ONLINE

3 River landscapes and processes

Physical processes shape river landscapes

What are the physical processes at work in rivers?

REVISED

What are the main types of weathering?

Mechanical, biological and chemical weathering have an impact on river landscapes. Look back at Chapter 2, page 12 to revise these processes.

Now test yourself

TESTED

Explain the process of biological weathering.

What is mass movement?

Sliding and slumping

- An area of the bank slips into the river.
- Due to the nature of the slip, it leaves behind a curved surface.
- During dry weather the clay contracts and cracks.
- When it rains, the water runs into the cracks and is absorbed until the rock becomes saturated.
- This weakens the rock and, due to the pull of gravity, it slips down the slope on its slip plane.

How are rivers affected by erosion, transportation and deposition?

The water in a river is continually attacking its bed and banks in a process known as erosion. The key terms box contains information on river processes.

Weathering is the way that rocks are broken down in situ.

Mass movement is when material moves down a slope, pulled down by gravity.

Transportation is the movement of materials by water.

Traction is the rolling of large sediment such as pebbles along the river bed.

Saltation occurs when small pieces of shingle or large grains of sand are bounced along the river bed.

Suspension is when small particles such as sand and clays are carried in the water. This can make the water look cloudy especially after heavy rainfall or when the river has lots of energy.

Hydraulic action is the pressure of water against the banks and bed of the river. It also includes the compression of air in cracks: as the water gets into cracks in the rock, it compresses the air in the cracks; this puts even more pressure on the cracks and pieces of rock may break off.

Abrasion occurs when particles being carried by the river are thrown against the river banks.

Solution is a chemical reaction between certain rock types and the river water.

Attrition is the breaking up of sediment being carried by the river. Stones and pebbles are continually knocked against each other by the water causing them to become smoother and smaller.

Deposition is when a river drops some of the sediment that it is carrying.

Now test yourself

Describe **three** processes used by rivers to erode their bed and banks.

TESTED

How do river landscapes contrast between the upper course, the mid course and the lower course?

REVISED ☐

As a river moves from upper to lower course, the following changes occur:

River characteristic	Change
Width	widens
Depth	deepens
Velocity	increases
Discharge	increases
Gradient	decreases
Volume	increases
Sediment size	decreases

Figure 1 **The changes in a river's course**

Source is the start of a river.

Mouth is where the river ends, either when it joins another river or meets the sea.

Long profile is a slice through the river from source to mouth which shows the changes in height of the river's course.

Valley profile is a slice across a river showing the changes in height across the valley.

Channel shape is the width and depth of the river.

Velocity is the speed of the river.

Discharge is the amount of water passing a specific point at a given time and is measured in cubic metres per second.

Gradient is the slope at which the river loses height.

Volume is the amount of water in the river.

Why do river characteristics change along the course of the River Creedy?

REVISED ☐

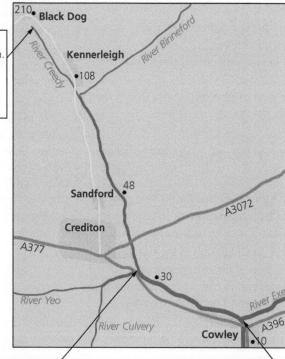

Its source is in the hills to the north of Crediton. In the upper course the river is shallow and narrow with a steep gradient.

The river becomes wider and deeper as it gains more water from other rivers. To the south-east of Crediton, it is met by the River Yeo adding more water to the river. This gives the river more power to erode.

The river is also moving away from the hills into flatter areas so the gradient of the river is becoming less and the river is less likely to erode vertically and more likely to erode laterally forming meander bends. The river ends at its confluence with the River Exe just north of Exeter.

Figure 2 **River characteristics of the River Creedy in Devon**

Now test yourself

Describe how river characteristics change as a river flows from its source to its mouth.

TESTED ☐

How does the UK's weather and climate affect river processes and impact on landforms and landscapes?

- Seasonality affects the rate at which the river can erode.
- Storms or heavy rainfall give the river greater erosional power. Riverbank trees can be undermined.

- In droughts rivers have less power.
- River defences are being installed and repaired more often due to greater storm frequency.

River erosion and deposition interact with geology to create distinctive landforms within river landscapes

What landforms are created by river erosion interacting with the geology of an area?

Interlocking spurs

Interlocking spurs are barriers of hard resistant rock, which the river cannot easily erode. The river weaves its way around them.

> **Exam tip**
>
> When answering questions about landform formation:
> - always include an explanation of processes
> - describe in detail and explain what happens in a sequence of events
> - drawing a diagram will help your answer.

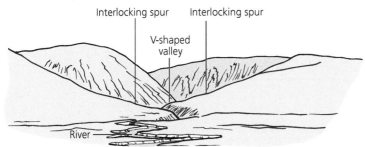

Figure 3 **Interlocking spur formation**

Waterfalls and gorges

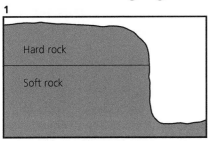

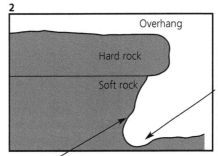

Figure 4 **Waterfall and gorge formation**

What landforms are created by river erosion and deposition?

Meander bends and oxbow lakes

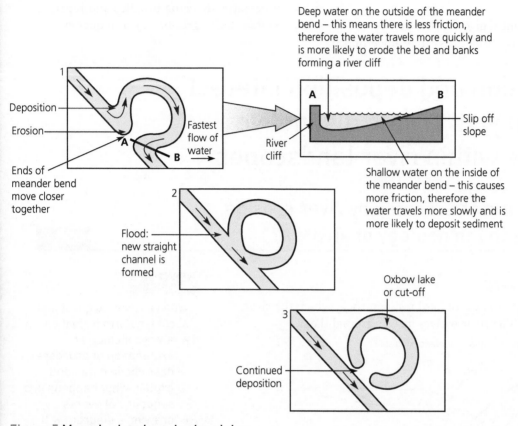

Deep water on the outside of the meander bend – this means there is less friction, therefore the water travels more quickly and is more likely to erode the bed and banks forming a river cliff

Deposition

Erosion

Ends of meander bend move closer together

Fastest flow of water

River cliff

Slip off slope

Shallow water on the inside of the meander bend – this causes more friction, therefore the water travels more slowly and is more likely to deposit sediment

Flood: new straight channel is formed

Oxbow lake or cut-off

Continued deposition

Figure 5 Meander bends and oxbow lakes

Now test yourself

Explain the formation of oxbow lakes.

What landforms are created by deposition?

Floodplain

- This is a low flat area of land on either side of a river.
- It is formed by the migration of meanders downstream.
- Lateral erosion causes meander bends to move across and down the valley in the direction of the river's flow.
- The outside of the bend, where erosion is greatest, moves the bend in that direction.
- The inside of the bend fills in the floodplain with the deposition that occurs there.

- At times of high river flows, this area will flood as the water moves out of the river channel onto the land that surrounds it.
- The water is shallower on the land than in the river channel. Therefore, there is more friction and the water drops the sediment it is carrying.
- The water drops the heaviest material first on the banks; the lighter material such as silt is carried the furthest. The deposit of this material forms a floodplain.

Levee

- This is a high bank at the side of a river which is built up during times of flood.
- Each time a river floods, it deposits sediment on its banks. This is because of the change in speed of the water between when it is in the channel and when it has moved out onto the floodplain.
- Over time these banks build up to form levees, which make it harder for the river to flood.

Figure 6 **Floodplain of Nant Ffrancon valley**

Exam practice

Examine how physical processes work together in the formation of the floodplain shown in Figure 6. (8 marks)

ONLINE

Human activities can lead to changes in river landscapes which affect people and the environment

How have human activities affected river landscapes?

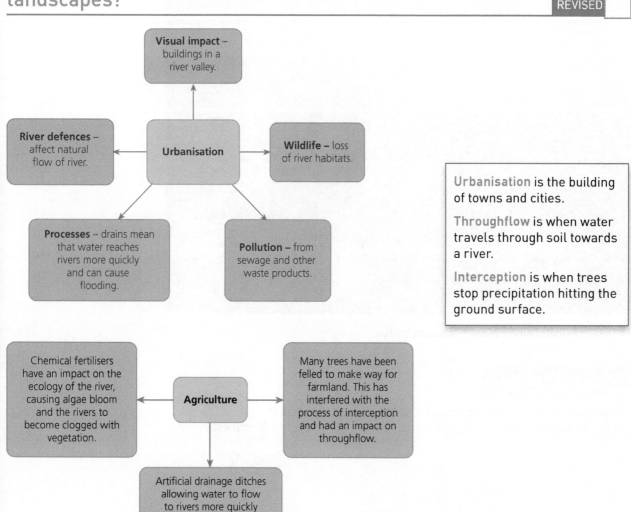

Visual impact – buildings in a river valley.

River defences – affect natural flow of river.

Urbanisation

Wildlife – loss of river habitats.

Processes – drains mean that water reaches rivers more quickly and can cause flooding.

Pollution – from sewage and other waste products.

Urbanisation is the building of towns and cities.

Throughflow is when water travels through soil towards a river.

Interception is when trees stop precipitation hitting the ground surface.

Chemical fertilisers have an impact on the ecology of the river, causing algae bloom and the rivers to become clogged with vegetation.

Agriculture

Many trees have been felled to make way for farmland. This has interfered with the process of interception and had an impact on throughflow.

Artificial drainage ditches allowing water to flow to rivers more quickly can cause flooding.

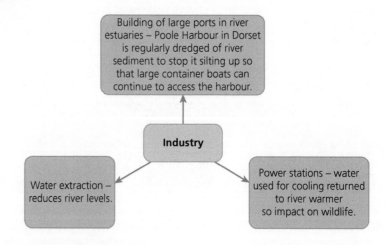

Building of large ports in river estuaries – Poole Harbour in Dorset is regularly dredged of river sediment to stop it silting up so that large container boats can continue to access the harbour.

Industry

Water extraction – reduces river levels.

Power stations – water used for cooling returned to river warmer so impact on wildlife.

What are the physical and human causes of river flooding?

Physical causes	Human causes
Heavy rainfall – if there are large amounts of rain day after day, the water will saturate the ground and flow more quickly into the river.	**Removal of vegetation on valley slopes** – then there is less interception and water will move to the river more quickly.
Cloudburst in a thunderstorm – the rain droplets are so large and fall so quickly that there is no time for the water to sink into the ground. Water runs very quickly into the river and causes flooding.	**Settlements built on the floodplain** – storm drains will allow water to move into the river at a greater speed and so make flooding more likely.
Sudden rise in temperature – a rapid thaw can happen. Rivers are unable to cope with the amount of water and flood.	**Global warming** – melting of polar ice caps and a rise in sea levels, flooding low-lying coastal areas.
Silted up river channels – make the channel smaller and more likely to flood.	**Dams may burst** – which causes excess water in river channels and flooding of large areas.

Figure 7 **The physical and human causes of river flooding**

Exam tip

In an exam question, you may be asked for just physical factors or just human factors, so be sure that you know which are which.

Physical causes are any occurrence that is natural.

Human causes are any occurrence that is created by humans.

Hydrograph is a graph showing rainfall and river discharge over a specific period of time.

Exam practice

Explain how deforestation can cause rivers to flood. (3 marks)

ONLINE

What are the effects of river flooding on people and the environment?

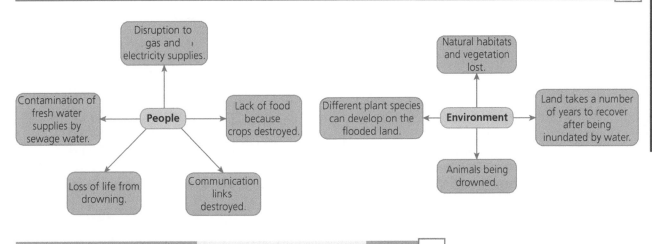

Now test yourself

TESTED

Using examples, explain the effects of river flooding.

What are the main types of hard and soft engineering used on UK rivers?

A number of different types of defences are used to manage rivers, but what are their advantages and disadvantages and how do they change river landscapes? River defences can be classified as either hard or soft engineering techniques.

> **Soft engineering** is a method of river management which works or attempts to work with the natural processes occurring. They tend to be visually unobtrusive. It does not tend to involve major construction work, for example washlands.
>
> **Hard engineering** is a method of river management which involves major construction work, for example dams.

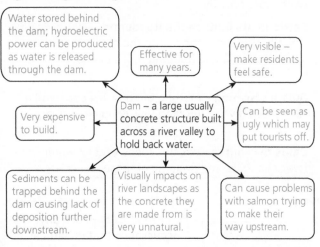

- Water stored behind the dam; hydroelectric power can be produced as water is released through the dam.
- Effective for many years.
- Very visible – make residents feel safe.
- Very expensive to build.
- Can be seen as ugly which may put tourists off.
- **Dam** – a large usually concrete structure built across a river valley to hold back water.
- Sediments can be trapped behind the dam causing lack of deposition further downstream.
- Visually impacts on river landscapes as the concrete they are made from is very unnatural.
- Can cause problems with salmon trying to make their way upstream.

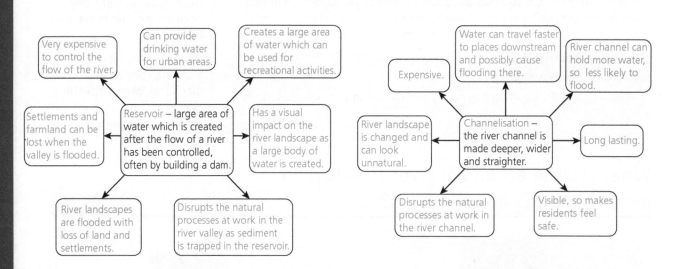

- Very expensive to control the flow of the river.
- Can provide drinking water for urban areas.
- Creates a large area of water which can be used for recreational activities.
- Settlements and farmland can be lost when the valley is flooded.
- **Reservoir** – large area of water which is created after the flow of a river has been controlled, often by building a dam.
- Has a visual impact on the river landscape as a large body of water is created.
- River landscapes are flooded with loss of land and settlements.
- Disrupts the natural processes at work in the river valley as sediment is trapped in the reservoir.

- Water can travel faster to places downstream and possibly cause flooding there.
- River channel can hold more water, so less likely to flood.
- Expensive.
- **Channelisation** – the river channel is made deeper, wider and straighter.
- Long lasting.
- River landscape is changed and can look unnatural.
- Disrupts the natural processes at work in the river channel.
- Visible, so makes residents feel safe.

- Large areas of land cannot be built on; residents may not understand why.
- Provide potential habitat for birds and animals.
- Provide recreational facilities such as sports fields for local residents
- Flooding takes the land out of action regularly; this would anger people who are using it for recreation.
- **Floodplain zoning** – land that is close to the river is seen as low value because of flood risk so usage is for recreation, often sports fields. Housing areas would be further away on more valuable land which is less likely to flood. **Washlands** – the river is allowed to flood these areas; could be farmland or recreational land close to settlements.
- Very cheap as no defences need to be built.
- Ecology of the landscape is changed each time the river floods.
- River landscapes are left relatively unchanged for the majority of the time.

Now test yourself

What are the differences between hard and soft engineering techniques?

Exam tip

Questions could be based on the advantages and disadvantages of one of these techniques or could be more general about hard and soft engineering techniques.

Interaction between physical and human processes leads to distinctive river landscapes

Located example: Lower Wye Valley between Tintern and Chepstow

Significance of the location

- The location forms part of the border between England and Wales.
- It has spectacular limestone scenery.
- The beauty of the area led to the birth of tourism. Many of the viewpoints that exist today were built for tourists in the mid-eighteenth century.

Physical or human process	Impact on (change to) the river landscape
Industry	Quarrying – the sides of the gorge have been extensively quarried for limestone. This was for building materials and limekilns. This has increased the slopes of the gorge. Iron ore smelting – the valley had a plentiful supply of water, iron ore and wood for charcoal. It was the perfect setting for early iron smelting in Britain.
River erosion	The river erodes and deposits material, forming meanders and floodplains (see page 24).
Weathering	The processes of mechanical, chemical and biological weathering are all present in the area, providing material for the river to use in erosion and deposition processes.
Forestry	Many trees were felled in the eighteenth and nineteenth centuries for ship building and other industrial uses such as making charcoal. Up to the Second World War, the woodlands were mainly deciduous. After this time, extensive planting led to the area having 40% of all its woods either dominated by conifers or a substantial amount of conifers and a few broadleaf trees. Since the 1980s, this planting has stopped and broadleaved trees are now the main type being planted. Some woodlands were destroyed completely. Others have appeared such as on Coppet Hill where a wood has replaced previous open common pasture.
Human development	A road was built along the valley in the early nineteenth century and the railway followed in 1876. Before this, the river was the economic backbone of the area, allowing access for industry and tourists. Settlement in the valley goes back 12,000 years. Offa's Dyke on the east bank of the river was built in the eighth century.
Tourism	The Wye Valley was one of the earliest tourist honeypots, with visitors flocking to the area in the 1700s. It was at this time that the cliff ascent and walks at Piercefield Park were landscaped. Tourists still flock to the area; there are many look-out points and walks. There are also a number of castles in the area, and Tintern Abbey which dates back to the eleventh century.

Figure 8 **Physical and human processes that have shaped the landscape of the Wye Valley**

Now test yourself

TESTED

Why is the Wye Valley a significant river landscape?

Exam practice

For a named example, examine how the distinctive river landscape is the outcome of physical and human processes.

(8 marks)

ONLINE

4 Glaciated upland landscapes and processes

Physical processes shape glaciated upland landscapes

What are the main types of glacial erosion?

REVISED

Ways in which glacial landscapes are eroded

Abrasion – as the glacier moves, the rocks which are in the glacier erode the sides and bottom of the valley.

Plucking – ice from the glacier melts due to pressure and the water runs into cracks on the mountain side. It refreezes almost immediately, breaking off pieces of rock, which become embedded in the glacier.

Supraglacial debris is rock material that is carried on the surface of a glacier.

Englacial debris is rock material that is carried in the main body of the glacier.

Drift refers to all sediments deposited by glaciers.

Till is material deposited directly by a glacier.

Fluvio-glacial material consists of rocks and other debris deposited by melting ice or glacial streams.

Exam practice

Explain the process of plucking.   (2 marks)

ONLINE

Exam tip

Remember to use these terms when explaining how glacial landforms are formed.

How do glaciers transport materials (rock debris)?

REVISED

Within the glacier – there is material in the ice of the glacier which could have been on the surface of the glacier but was buried by snow falls.

Ways in which glaciers transport material

On the surface of the glacier – a glacier carries material on its surface, at the sides close to the valley slopes (lateral moraine) and in the middle (medial moraine).

How do glaciers deposit materials?

REVISED

- If there is a change in power of the glacier, it moves more slowly due to increased friction and material (such as moraine) will be deposited.
- Glacial streams under the glacier or flowing out of the glacier deposit materials.

Now test yourself

Explain how a glacier deposits material.

TESTED

What processes are occurring in upland glacial landscapes of today?

Mechanical weathering or freeze-thaw action

A small crack in a rock fills with water during the daytime. As the water begins to freeze at night, it starts at the top, sealing the crack.

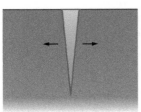

As the water freezes completely, its 9% growth exerts an outward force on the sides of the crack, increasing the size of the crack by a maximum of 9%.

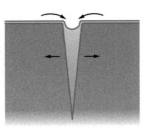

If the ice thaws the next day the resulting water will not fill the crack, which is now both wider and deeper because of its 9% expansion.
Dew or rainfall on the rock surface can refill the crack.

The process begins again, this time with a larger initial crack.

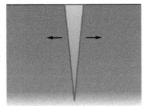

Again the crack expands by as much as 9%. Continued freezing and thawing, particularly with the daily addition of the water to keep the crack full, eventually leads to significant fracturing of the rock.

Figure 1 **The process of freeze-thaw**

Mass movement

Material moves down a slope due to the pull of gravity. This includes soil movement such as slumping and soil creep. Soil creep is when gravity very slowly pulls the water that is in soil down a slope; the soil particles move slowly down the slope with the water. Unlike slumping it is not possible to see this happening, although the slope may appear rippled (like sheep paths around a hill). Mass movement also includes rock falls, which occur often in glaciated upland areas.

Figure 2 **Glaciated upland area**

How do past climate and current UK weather and climate affect processes that impact on glaciated upland landscapes?

REVISED

Glaciated landscapes were formed during the last ice age about 20,000 years ago. As the climate got colder, more precipitation fell as snow. The climate continued to get colder until the snow did not melt and over a number of years became compacted into ice – over a kilometre thick in Scotland!

The seasonality of the UK's present weather and climate will affect the processes at work in glaciated areas. In winter, snow accumulates in mountainous areas such as Snowdonia but not enough to greatly change the landscape. Another process at work on the landscape is mechanical weathering which will be more active if there are great changes in temperature between day and night. The landscape is also constantly being changed by rivers and the sea.

Glacial erosion and deposition create distinctive landforms

What landforms are created by glacial erosion?

REVISED

> **Stoss end** is the side of a landform facing the direction of the ice flow.
>
> **Lee slope** is the side of a landform facing away from the direction of the ice flow.
>
> **Volcanic plug** is a volcanic cone from an extinct volcano.

Corries and arêtes	Glacial troughs and truncated spurs
<ul><li>Snow is compacted into ice on mountainsides.</li><li>Ice starts to move due to the pull of gravity.</li><li>Plucking forms the steep back wall.</li><li>Abrasion shapes the bottom of the hollow.</li><li>Erosion rates are slower here due to less pressure.</li><li>Rock lip forms; size increases due to deposition of rock material as the ice slows down to leave the corrie.</li><li>After glaciation, the corrie fills with water kept in by this rock lip.</li><li>If two corries form beside each other on a mountainside their sides will become knife-edge ridges known as arêtes. These are continually attacked by freeze-thaw weathering.</li></ul>	<ul><li>A valley glacier will completely fill a river valley. It has a lot of power to erode because of its size and weight.</li><li>This great power cuts back the interlocking spurs using abrasion and plucking, forming truncated spurs.</li><li>This leaves a wide, flat-bottomed valley with steep cliff-like sides.</li></ul>
Roche moutonnées	**Hanging valleys**
<ul><li>These range from small rocky outcrops a few metres high to small hills a few 100 metres high.</li><li>They are formed from an outcrop of more resistant rock which was in the path of the ice as it moved across the upland area.</li><li>The ice slows as it moves up the hard rock and more melting is occurring. On this stoss end side, abrasion will take place causing a smooth upstream slope to form.</li><li>On the lee slope, the ice is under less pressure and refreezes around the rock, pulling rock fragments away as it moves (plucking).</li></ul>	<ul><li>These are the valleys of tributary streams.</li><li>They are left high up on the valley sides of the main valley because their glaciers were smaller and had less erosive power than the main valley glaciers.</li><li>When the ice melted, these valleys were at a higher level.</li><li>These streams now meet the main river via a waterfall.</li></ul>

Figure 3 **Landforms created by glacial erosion**

Figure 4 **A corrie in Snowdonia**

What landforms are created by glacial deposition?

REVISED

Ground and terminal moraine

- This is material transported and deposited by glaciers.
- Moraine is collected by the glacier using freeze-thaw weathering, plucking and abrasion.
- Moraine is dropped where the glacier slows down.
- Ground moraine is deposited below the glacier.
- Terminal moraine is dropped at the end of the glacier.

What landforms are created by glacial erosion and deposition?

REVISED

Crag and tail	Drumlins
• Crags form when a glacier meets a hard, resistant rock such as granite or a volcanic plug. • The glacier will pass over the hard rock but leave it standing, eroding the softer rock in front of it. • On the lee slope, material is deposited. This appears as a ramp going away from the crag in the direction of the ice movement.	• Drumlins are elongated landforms which are formed of till. • They can be a kilometre or more in length, 500 m in width and 50 m in height. • The slope facing the ice flow or stoss end is steeper than the lee slope which tapers away to ground level. • Drumlins are often found in groups. A collection of drumlins together is called a swarm. • Drumlins were formed when the ice was carrying a lot of material. If the ice had to slow down due to an obstruction in its path, it would deposit some of the material it was carrying. • Due to a change in speed, most material would be deposited at this point and then gradually less as the ice continued on its path.

Figure 5 **Landforms created by glacial erosion and deposition**

Human activities can lead to changes in glaciated upland landscapes

How have human activities such as farming, forestry and settlement affected glaciated upland landscapes?

Settlements developed originally as farming communities and market towns but there has now been further development as tourism in upland areas continues to grow. Skiing resorts have been built in mountainous areas such as the Alps but also in the UK in the Cairngorms in Scotland. These settlements have a visual impact on upland areas and development has also had an impact on the ecosystems of these areas.

Farming has had an impact on glaciated areas. Originally many of the UK's glaciated areas would have been wooded but they are now moorlands with rough pasture. For many years, sheep farming was the main use of the land but this is changing with the demand for venison (deer meat). There will soon be more deer on the Island of Arran in Scotland than sheep because farmers can earn more money from game shooting and venison.

Forestry has also had an impact on upland areas. As mentioned earlier, much of the UK's upland areas were once woodland. The wood was felled for fire wood but also for shipbuilding and many other uses. In the latter half of the twentieth century, native woodland was replaced with fast-growing evergreens. Now, however, the forestry is much more in keeping with the landscape, being a mix of coniferous and deciduous woodland.

> **Coniferous woodland** is an area of trees which do not lose their leaves in winter (evergreens).
>
> **Deciduous woodland** is an area of trees which lose their leaves in winter.
>
> **SSSI** is a site of special scientific interest. It is protected for its unique habitat.
>
> **HEP** is hydroelectric power.
>
> **Onshore windfarm** is an area of land which is covered by wind turbines.

Now test yourself

TESTED

Describe how farming and forestry have changed upland areas.

What are the advantages and disadvantages of development in glaciated upland landscapes?

Glaciated upland areas have been developed in a number of ways. This development can have both advantages and disadvantages for the local area. It can also lead to changes to the glaciated landscape.

Water storage and supply

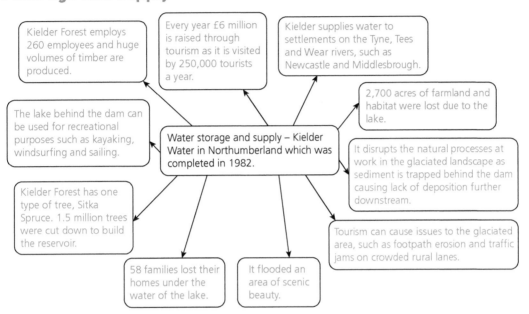

Kielder Forest employs 260 employees and huge volumes of timber are produced.

Every year £6 million is raised through tourism as it is visited by 250,000 tourists a year.

Kielder supplies water to settlements on the Tyne, Tees and Wear rivers, such as Newcastle and Middlesbrough.

The lake behind the dam can be used for recreational purposes such as kayaking, windsurfing and sailing.

Water storage and supply – Kielder Water in Northumberland which was completed in 1982.

2,700 acres of farmland and habitat were lost due to the lake.

It disrupts the natural processes at work in the glaciated landscape as sediment is trapped behind the dam causing lack of deposition further downstream.

Kielder Forest has one type of tree, Sitka Spruce. 1.5 million trees were cut down to build the reservoir.

Tourism can cause issues to the glaciated area, such as footpath erosion and traffic jams on crowded rural lanes.

58 families lost their homes under the water of the lake.

It flooded an area of scenic beauty.

Renewable energy

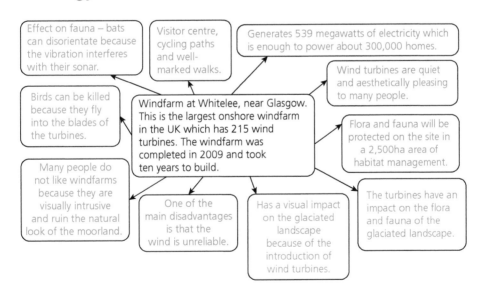

Effect on fauna – bats can disorientate because the vibration interferes with their sonar.

Visitor centre, cycling paths and well-marked walks.

Generates 539 megawatts of electricity which is enough to power about 300,000 homes.

Birds can be killed because they fly into the blades of the turbines.

Windfarm at Whitelee, near Glasgow. This is the largest onshore windfarm in the UK which has 215 wind turbines. The windfarm was completed in 2009 and took ten years to build.

Wind turbines are quiet and aesthetically pleasing to many people.

Flora and fauna will be protected on the site in a 2,500ha area of habitat management.

Many people do not like windfarms because they are visually intrusive and ruin the natural look of the moorland.

One of the main disadvantages is that the wind is unreliable.

Has a visual impact on the glaciated landscape because of the introduction of wind turbines.

The turbines have an impact on the flora and fauna of the glaciated landscape.

> **Exam tip**
>
> Questions could be based on the advantages and disadvantages of one of these developments or could be about the advantages and disadvantages of development in general.

Recreation and tourism

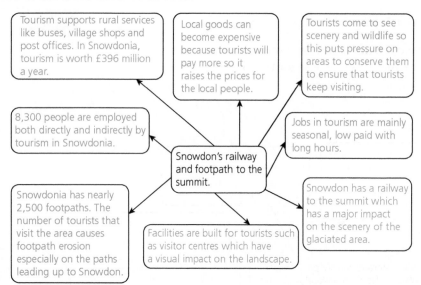

Tourism supports rural services like buses, village shops and post offices. In Snowdonia, tourism is worth £396 million a year.

Local goods can become expensive because tourists will pay more so it raises the prices for the local people.

Tourists come to see scenery and wildlife so this puts pressure on areas to conserve them to ensure that tourists keep visiting.

8,300 people are employed both directly and indirectly by tourism in Snowdonia.

Jobs in tourism are mainly seasonal, low paid with long hours.

Snowdon's railway and footpath to the summit.

Snowdonia has nearly 2,500 footpaths. The number of tourists that visit the area causes footpath erosion especially on the paths leading up to Snowdon.

Facilities are built for tourists such as visitor centres which have a visual impact on the landscape.

Snowdon has a railway to the summit which has a major impact on the scenery of the glaciated area.

Conservation

Its main aims are to stop the spread of non-native species, to maintain and repair footpaths on popular routes and to co-ordinate voluntary conservation efforts. There are also a number of planning laws which are there to conserve the environment but can have advantages and disadvantages.

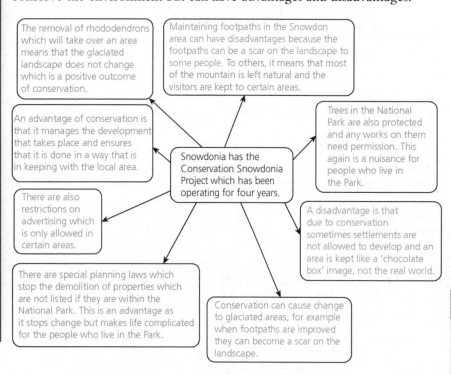

The removal of rhododendrons which will take over an area means that the glaciated landscape does not change which is a positive outcome of conservation.

Maintaining footpaths in the Snowdon area can have disadvantages because the footpaths can be a scar on the landscape to some people. To others, it means that most of the mountain is left natural and the visitors are kept to certain areas.

An advantage of conservation is that it manages the development that takes place and ensures that it is done in a way that is in keeping with the local area.

Trees in the National Park are also protected and any works on them need permission. This again is a nuisance for people who live in the Park.

Snowdonia has the Conservation Snowdonia Project which has been operating for four years.

There are also restrictions on advertising which is only allowed in certain areas.

A disadvantage is that due to conservation sometimes settlements are not allowed to develop and an area is kept like a 'chocolate box' image, not the real world.

There are special planning laws which stop the demolition of properties which are not listed if they are within the National Park. This is an advantage as it stops change but makes life complicated for the people who live in the Park.

Conservation can cause change to glaciated areas, for example when footpaths are improved they can become a scar on the landscape.

Now test yourself

Draw a table which shows the advantages and disadvantages of development in upland areas.

TESTED

Exam practice

Renewable energy has both negative and positive effects on glaciated landscapes.

Explain **two** ways that renewable energy has a negative effect on glaciated landscapes. [4 marks]

ONLINE

Exam tip

Read the question carefully and don't get caught out. Underline the key words in the question.

Interaction between physical and human processes leads to distinctive glaciated upland landscapes

Located example: Isle of Arran

Significance of the location

- The location is significant because the island is cut in two by the highland boundary fault which is the line between the highlands and lowlands of Scotland. The island is like Scotland in miniature.
- The north of the island is a rugged mountainous area created from a large granite batholith about 60 million years ago. This large mass of molten magma pushed its way up into the Earth's crust, lifting rocks above it by about 3,000 metres. As these rocks were eroded by ice and water, the granite underneath was exposed. The area was then carved into its distinctive landscape by physical processes during the last ice age.
- The south of the island has more gentle slopes with farmland and forests.

Physical and human processes that have shaped the landscape of the Isle of Arran

Physical or human process	Impact on (change to) the river landscape
Farming: the impact of sheep and more recently deer farming has changed the landscape; there are over 2,000 red deer in the north of the island. This has caused a loss of native habitats which are sensitive to overgrazing.	**Freeze-thaw weathering:** this continues to have an impact on the glaciated landscape. Scree slopes are common as is rock debris from arêtes as they are continually sharpened by freeze-thaw weathering.
Tourism: there are seven golf courses on Arran, the island is promoted as a golfing venue. This has changed the glaciated landscape as the links and fairways are not a native habitat for the island. Footpath erosion is also a problem on the island. Many of the local businesses charge extra for certain services and give the money to the Arran Trust which is a charity that works to protect the environment. There is also the problem of litter and disturbance to livestock.	**River processes:** the rivers on Arran continue to erode the glaciated landscape. There are a number of river features such as meander bends which are present in the glacial troughs. Waterfalls and interlocking spurs can also be seen. These are changes that have occurred due to the work of rivers on the glaciated landscape.
Forestry: Arran was originally covered in woodlands and moorlands. A quarter of the island is now covered in forests, mostly coniferous plantations. This has had an impact on the landscape because the flora and fauna of these forests are different to those of the traditional deciduous forests and moorland.	The processes of mechanical, chemical and biological weathering are all present in the area, providing material for the river to use in erosion and deposition processes.

Figure 6 **Physical and human processes that have shaped the Isle of Arran**

Exam practice

For a named example, examine how the distinctive glaciated upland landscape is the outcome of physical and human processes.

(8 marks)

ONLINE

5 Weather hazards and climate change

The atmosphere operates as a global system transferring heat and energy

> **Global atmospheric circulation** is the worldwide movement of air which transports heat from tropical to polar latitudes.
>
> **Ocean current** is a continuous, directed movement of ocean water. The currents are made from forces acting on the water such as the wind, different temperatures and the Earth's rotation.
>
> **Hemisphere** is a half of the Earth. The northern hemisphere is above the Equator, the southern hemisphere is below the Equator.
>
> **ITCZ** is the Inter Tropical Convergence Zone.
>
> **Troposphere** is the lowest layer of the atmosphere. It is thicker at the Equator (approximately 20 km) than at the poles (approximately 10 km).
>
> **Depression** is a low-pressure system which produces clouds, wind and rain.
>
> **Atmosphere** consists of the gases that surround the Earth.

What are the features of the global atmospheric circulation?

REVISED

The features of the global atmospheric circulation are:
- The transfer of heat from the Equator to the poles.
- There are three circulation cells – Hadley, Ferrel and Polar.
- Jet streams impact on the movement of heat energy.
- The spin of the Earth creates the Coriolis effect.

How do circulation cells and ocean currents transfer and redistribute heat energy across the Earth?

REVISED

The main source of heat energy for the world is the Sun. The Sun heats the Earth's surface unevenly. It heats the Earth more at the Equator than at the poles. This creates a heat surplus at the Equator and a heat deficit at the poles. As the poles are not getting colder and the Equator is not getting noticeably warmer, there must be a redistribution of heat energy across the Earth. But how does this work?

The heat energy is transferred in two ways: circulation cells and ocean currents.

Circulation cells	Features
Hadley cells stretch from the equator to latitudes 30°N and 30°S.	• Warm trade winds blow towards the Equator. • At the Equator, the trade winds from each hemisphere meet. The warm air rises rapidly causing thunderstorms. An area of low pressure is formed in the ITCZ where the air from the two cells meets over the Equator. • The air at the top of the troposphere moves towards 30°N and 30°S where it becomes cooler and starts to sink back to the Earth's surface. As it descends, it warms and any moisture is evaporated. • This creates high-pressure areas, with cloudless skies. The world's hot deserts are found in these areas, such as the Sahara, in North Africa. • On returning to the ground, some of the air returns to the equatorial areas as trade winds; this completes the circle.
Ferrel cells stretch from latitudes 30°N and 30°S to latitudes 60°N and 60°S.	• Air on the surface is pulled towards the poles. This forms the warm southwesterly winds in the northern hemisphere and northwesterly winds in the southern hemisphere. • These winds collect moisture as they blow over oceans on the Earth's surface. • At about 60°N and 60°S they meet cold air from the poles. • The warm air rises over the cold air as it is less dense. This produces low pressure at the Earth's surface and pressure systems known as depressions. • Some of the air returns to the tropics and some is diverted to the poles as part of the polar cell. • The cell has a motion to the right in the northern hemisphere and to the left in the southern hemisphere due to the spin of the Earth. This is called the Coriolis effect.
Polar cells stretch from latitudes 60°N and 60°S to the north and south poles.	• The air sinks over the poles producing high pressure. • The air then flows towards the low pressure in the mid-latitudes, about 60°N and 60°S. Here it meets the warm air of the Ferrel cells.

Figure 1 **Features of circulation cells**

The polar front jet stream is formed when cold polar air meets warm tropical air high above the Atlantic Ocean, usually between latitudes 40° and 60°N and 40° and 60°S. It moves in a westerly direction. It marks the division between the Polar and Ferrel cells.

Jet streams are bands of extremely fast-moving air in the upper atmosphere.

The subtropical jet stream is generally in a westerly direction. It can be found at approximately 25°N and 35°S.

Now test yourself

Explain how circulation cells and ocean currents redistribute heat energy across the Earth.

TESTED ☐

Ocean currents

- The oceans transfer approximately 20 per cent of the total heat that is transferred from the tropics to the poles.
- Each ocean has a circular pattern of surface currents, known as a gyre.
- They are produced as masses of water move from one climatic zone to another.
- They are created by the surface winds generated by the global atmospheric circulation.

- In the northern hemisphere, currents move in a clockwise direction and in the southern hemisphere they move in an anticlockwise direction.
- The strongest currents are on the western side of oceans. For example, warm ocean currents such as the North Atlantic Drift transfers heat from low to high latitudes. This is particularly noticeable between latitudes 40° and 65° in winter.

The global climate continues to change due to natural causes

How has climate changed in the past over different time scales: glacial and interglacial periods during the Quaternary period?

- The Quaternary period is the past 1.8 million years of the world's history.
- During this period, there have been times when the temperature of the world has dropped. These are known as glacial periods (or ice ages).
- The temperature then became warmer, melting the large ice sheets which had formed. These are known as interglacial periods.

- The most recent glacial period occurred between about 120,000 and 11,500 years ago. Since then, the Earth has been in an interglacial period called the Holocene epoch. The remnants of the last ice age still cover ten per cent of the Earth's surface in Greenland, Antarctica and mountainous regions.

What are the causes of natural climate change?

This is a change in the amount of heat energy that comes from the Sun. These variations are very small. Sunspots on the Sun's surface do seem to have an impact on the heat energy of the Sun and therefore the climate of the Earth. There was a reduction in sunspot activity between 1645 and 1715, which corresponds with the Little Ice Age. Since the 1940s there has been a lot of sunspot activity, which could be a reason for the Earth's climate becoming warmer.

Eccentricity – the path of the Earth's orbit around the sun is not a perfect circle – it is an ellipse. This shape can change from nearly circular to elliptical and back again. The measure of this change is called its eccentricity. It appears that cold periods occur when the Earth's orbit is more circular and warmer periods occur when it is more elliptical.
Precession – the Earth's axis wobbles like a spinning top. This cycle has an impact on the seasons and can cause warmer summers.
Axial tilt – the Earth is spinning on its own axis. The axis is not upright but tilts at an angle between 22.1° and 24.5°. The greater the degree of tilt is associated with the world having a higher temperature.

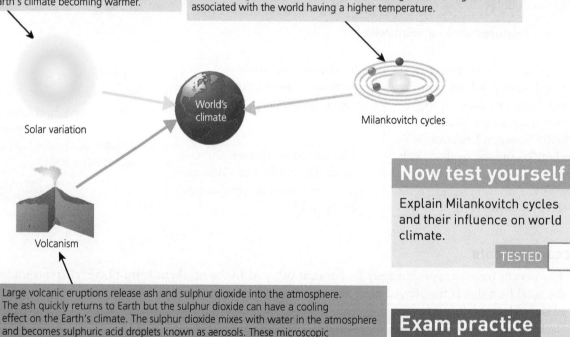

Solar variation

World's climate

Milankovitch cycles

Volcanism

Large volcanic eruptions release ash and sulphur dioxide into the atmosphere. The ash quickly returns to Earth but the sulphur dioxide can have a cooling effect on the Earth's climate. The sulphur dioxide mixes with water in the atmosphere and becomes sulphuric acid droplets known as aerosols. These microscopic droplets absorb radiation from the sun, heating themselves and the surrounding air. This stops heat reaching the Earth's surface. During the 1900s there were three large eruptions that may have caused the planet to cool down by as much as 1°. Eventually the effect will decrease as the aerosols fall as rain.

Figure 2 **The factors that affect the world's climate**

Now test yourself

Explain Milankovitch cycles and their influence on world climate.

Exam practice

Suggest how volcanoes can cause climate change. **(2 marks)**

What evidence is there for natural climate change?

Ice cores	Pollen records
The ice in areas such as Antarctica and Greenland has been there for millions of years. Cores can be drilled into it to measure the amount of carbon dioxide trapped in the ice. This is a climatic indicator because levels of carbon dioxide tend to be lower during cooler periods and higher when it is warmer.	Pollen analysis shows which plants were dominant at a particular time due to the climate. Each plant species has specific climatic requirements which influence their geographic distribution. All plants have a distinctive shape to their pollen grain. If these fall into areas such as peat bogs, they resist decay. Changes in the pollen found in different levels of the bog indicate changes in climate over time.
Tree rings	**Historical sources**
Each year the growth of a tree is shown by a single ring. If the ring is narrow, it indicates a cooler drier year. If it is thicker, it means the temperature was warmer and wetter. The patterns of growth are used to produce tree-ring timescales which give accurate climate information.	These include cave paintings, diaries and documentary evidence, for example, the fairs on the Thames when it froze. Since 1873, daily weather reports have been kept. Parish records are a good source of climate data for a particular area.

Figure 3 **Evidence for natural climate change**

Exam practice

State **two** pieces of evidence for natural climate change in the past.

(2 marks)

ONLINE

Exam tip

If you find it difficult to learn numbers, then learn names and places instead.

Global climate is changing as a result of human activity

How do human activities produce greenhouse gases that cause the enhanced greenhouse effect?

- **Industry**: some industries emit large amounts of greenhouse gases such as carbon dioxide and methane. This occurs during the production process, for example during the production of iron and steel, chemicals and cement.
- **Transport**: most forms of transport use fossil fuels to power them. When fossil fuels are burnt, gases such as carbon dioxide are released, which build up in the atmosphere adding to the enhanced greenhouse effect.
- **Energy**: the generation of power accounts for 25 per cent of global carbon dioxide emissions. The main source is the use of coal and natural gas to produce electricity.
- **Farming**: livestock, especially cattle, produce methane as part of their digestion. This represents almost one-third of the emissions from the agriculture sector. An increase in rice production due to growing populations in Asia has also seen an increase in the production of methane.

Fossil fuels are fuels which are produced from coal, oil and natural gas.

Enhanced greenhouse effect is also called climate change or global warming. It is the impact on the global climate of the increased amounts of carbon dioxide and other greenhouse gases that humans have released into the Earth's atmosphere since the Industrial Revolution.

Methane is a gas which comes from organisms that were alive many years ago, recently dead rotting organisms and those alive today.

Exam tip

It might be useful to know the sources of methane.

What are the negative effects that climate change is having on the environment and people?

Feature	Negative effect
Changing patterns of crop yield	In Africa, countries such as Tanzania and Mozambique will have longer periods of drought and shorter growing seasons. They could lose almost a third of their maize crop. It is forecast that in India there will be a 50% decrease in the amount of land available to grow wheat.
Rising sea levels	Research published in 2007 by the Met Office Hadley Centre for Climate Science and Services in Exeter, showed that between 1993 and 2006 sea levels rose 3.3 mm a year. This rise will threaten large areas of low-lying coastal land including major world cities such as London, New York and Tokyo.
Retreating glaciers	The vast majority of the world's glaciers are retreating (that is, melting), some more quickly than others. This is thought by some to be due to the increase in temperatures caused by climate change. The melting of the glaciers at the poles could also affect ocean water movement. It is believed that melting ice in the Arctic could cause the Gulf Stream to be diverted further south.

Figure 4 **Negative effects of climate change on the environment and people**

Now test yourself

State the negative effects of climate change.

The UK's distinct climate has changed over time

Climate is the average temperature and precipitation figures for an area.

Weather is the day-to-day changes in temperature and precipitation.

Precipitation is any form of moisture that reaches the Earth; rain, snow, etc.

Annual temperature range is the difference between the highest and lowest temperatures of a place.

Total annual rainfall is the sum of all the rainfall that falls in a year in an area.

Ice fairs were amusements held on the River Thames during the Little Ice Age.

Source region is a large area of the Earth's surface where the air has a uniform temperature and humidity.

Prevailing wind is the direction from which the wind usually blows. In the UK, it is the southwest.

Latitude is the distance north or south of the equator. It is measured in degrees with the maximum being 90°N or 90°S.

How has the UK climate changed over the last 1,000 years?

REVISED

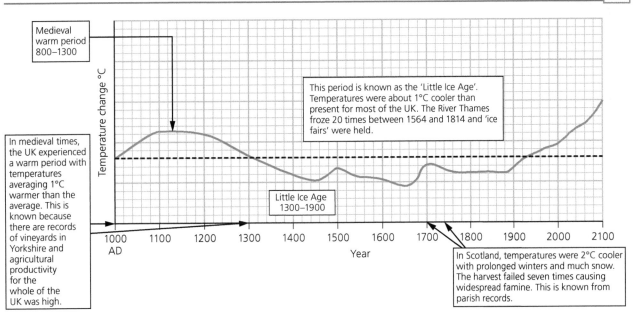

Medieval warm period 800–1300

In medieval times, the UK experienced a warm period with temperatures averaging 1°C warmer than the average. This is known because there are records of vineyards in Yorkshire and agricultural productivity for the whole of the UK was high.

This period is known as the 'Little Ice Age'. Temperatures were about 1°C cooler than present for most of the UK. The River Thames froze 20 times between 1564 and 1814 and 'ice fairs' were held.

Little Ice Age 1300–1900

In Scotland, temperatures were 2°C cooler with prolonged winters and much snow. The harvest failed seven times causing widespread famine. This is known from parish records.

Figure 5 **How the UK climate has changed over the past 1,000 years**

What is the UK climate like today and how does temperature, prevailing wind and rainfall vary within the UK?

REVISED

- The UK has a maritime climate with maximum average temperatures being 15 °C and minimum being 4°C, with gradual change.
- Rain falls every month; the total amount varies with location within the UK from approximately 550 mm in London to 1,800 mm in Fort William.
- There is little difference between the wettest and driest months.

- The prevailing wind of the UK is from the southwest and this varies little within the country. As this wind is blowing over the Atlantic Ocean, it will bring rainfall to the UK. The west of the country therefore receives more rainfall than the east.
- The temperature variations are the result of the influence of latitude and the distance the settlement is from the sea.

What is the significance of the UK's geographic location in relation to its climate?

Latitude

The latitude of the UK will impact on the amount of heat energy it receives from the Sun as places closer to the Equator are warmer than those at the poles. This is explained in Figure 6. Latitude also affects the temperature by influencing the length of the days. In the winter, the day length is short. This means that there are fewer hours of sunlight, resulting in lower temperatures.

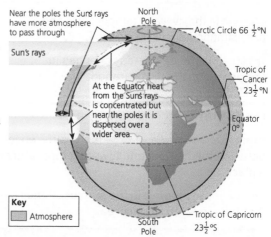

Figure 6 How latitude affects temperature

Air masses

The geographical location of the UK means that its climate is influenced by five air masses. This is unusual and helps to account for the changeable weather that is experienced by the UK. Each air mass has different weather characteristics (see Figure 7).

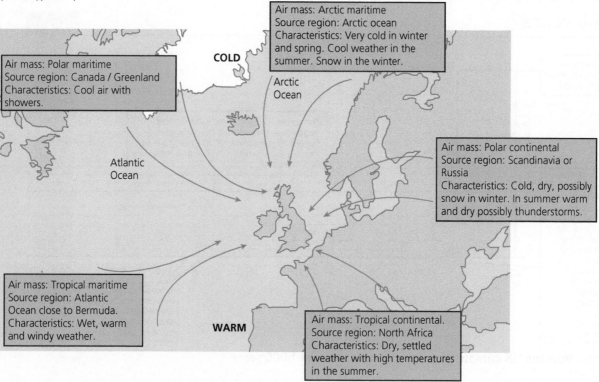

Air mass: Arctic maritime
Source region: Arctic ocean
Characteristics: Very cold in winter and spring. Cool weather in the summer. Snow in the winter.

Air mass: Polar maritime
Source region: Canada / Greenland
Characteristics: Cool air with showers.

Air mass: Polar continental
Source region: Scandinavia or Russia
Characteristics: Cold, dry, possibly snow in winter. In summer warm and dry possibly thunderstorms.

Air mass: Tropical maritime
Source region: Atlantic Ocean close to Bermuda.
Characteristics: Wet, warm and windy weather.

Air mass: Tropical continental.
Source region: North Africa
Characteristics: Dry, settled weather with high temperatures in the summer.

Figure 7 Air masses that affect the UK's weather

Distance from the sea

The distance a settlement is from the sea has an effect on its climate. Settlements that are close to the sea will have less extreme temperatures than places further inland. In the winter, settlements close to the sea will be warmer than settlements inland and in summer they will be cooler. The reason for this is that the land and sea respond differently to the heat energy from the Sun.

Ocean currents

Ocean currents also have an impact on temperatures. The UK is influenced by a warm current called the North Atlantic Drift. This current originates close to the Equator, moves through the Caribbean and across the Atlantic, and almost circles the UK. It is a warm body of water and significantly raises the temperature experienced by the UK.

Tropical cyclones are extreme weather events that develop under specific conditions and in certain locations

What are the characteristics, frequency and geographical distribution of tropical cyclones?

REVISED

Characteristics of tropical cyclones:
- They develop over tropical and sub-tropical oceans between the Tropic of Cancer and the Tropic of Capricorn with a water temperature of over 27°C.
- They usually form towards the end of the summer and in the autumn.
- They feature strong winds and heavy rain, often thunderstorms.
- The average wind speed is 120 kph but winds of 400 kph have been known.
- They normally move from east to west with the trade winds.
- They have an 'eye' which is the calm centre of the storm.
- Once they reach a wind speed of 60 kph, they are given a name, each letter of the alphabet being used in turn. The lists rotate every six years. If a storm is particularly destructive, the name is not used again and a new one is chosen. Floyd was replaced with Franklin in 2005.

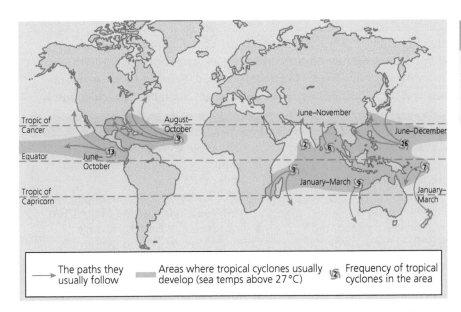

Now test yourself

State **four** characteristics of tropical cyclones.

TESTED

Figure 8 **The geographical distribution, frequency and characteristics of tropical cyclones**

What are the causes of tropical cyclones and what is the sequence of their formation?

REVISED

- Tropical cyclones all start over oceans with a minimum temperature of over 27°C.
- Hot air rises, taking a lot of water vapour with it.
- As it rises, it cools to form big cumulus clouds, creating low pressure at sea level.
- Air with higher pressure then moves in to replace it. This air does not move straight into the low-pressure area because of the Earth's circulation; it whirls into it. This air then moves upwards with more water vapour.
- This has two effects: storm clouds are pulled into a spin by the incoming wind and the spinning storm is pulled outward leaving a low-pressure funnel, the eye, in the centre. The cold air which is under high pressure sinks down into the centre, is heated and pulled into the spinning circle of air.
- The spinning circle begins to drift sideways because of the trade winds. This huge bundle of energy depresses the sea level under it, so there is a ridge of sea water giving rise to storm surges both before and after the cyclone has passed.

There are various impacts of and responses to natural hazards caused by tropical cyclones depending on a country's level of development

Why are tropical cyclones natural weather hazards?

Tropical cyclones are hazards because they cause:
- high winds
- intense rainfall
- storm surges
- coastal flooding
- landslides.

> **Exam tip**
>
> Don't try to learn all the specific case study information. Choose facts or figures, then learn one for each of the impacts. It is impossible to learn all of them. Many are given to give you a choice, not for you to learn all of them.

What were the social, economic and environmental impacts of Hurricane Sandy on Cuba and the USA?

On 23 October 2012, the government of Cuba warned the eastern states of the country about the approach of Hurricane Sandy. The hurricane continued north, affecting 24 states of the USA with particularly severe damage in New Jersey and New York. The hurricane had different social, economic and environmental impacts on Cuba and the USA because of the stage of development of each country.

Impacts	Cuba	USA
Social	<ul><li>There was no electricity or fresh water.</li><li>11 people were killed.</li><li>17,000 homes were destroyed, 226,000 were damaged.</li><li>55,000 people were evacuated because of the storm surge.</li></ul>	<ul><li>117 people were killed.</li><li>Roughly nine million homes had power cuts.</li><li>650,000 homes were damaged or destroyed in the whole of the USA.</li><li>The streets of New York were flooded, as was the subway.</li></ul>
Economic	<ul><li>Total losses in Santiago area came to £50 million.</li><li>Road to airport blocked so no tourists could arrive or leave the island, causing a loss in revenue.</li><li>US$2 billion (2012 USD) in total losses.</li></ul>	<ul><li>Insurance claims in New Jersey totalled US$3.3 billion.</li><li>US$1.1 billion was spent repairing the damage to sewage and water pipes in New Jersey and New York.</li><li>Damage cost in New York totalled US$19 billion.</li></ul>
Environmental	<ul><li>2,600 ha of banana crops were destroyed; some of the crop was close to Maisí town.</li><li>In Santiago de Cuba trees were uprooted and lost all of their leaves.</li><li>Coffee plantations in mountainous areas were swept away.</li><li>Areas close to the coast were flooded, with beaches being swept away, destroying wildlife habitats.</li></ul>	<ul><li>The storm surge meant that sea water got into fresh water habitats, having severe impacts on the wildlife.</li><li>1.5 billion litres of sewage were released into the Raritan River in New Jersey.</li><li>90% of beaches in New York and New Jersey were damaged.</li><li>1.5 million litres of oil were spilt into Arthur Kill, destroying wildlife habitats and killing fish and birds.</li></ul>

Figure 9 Social, economic and environmental impacts of Hurricane Sandy on Cuba and the USA

What were the different responses of individuals, organisations and governments in Cuba and the USA to Hurricane Sandy?

Responses by	Cuba	USA
Individuals	• After the hurricane, many people moved in with relatives or friends. • They used materials provided by the government and other organisations to rebuild their own homes. • The people of Cuba have no home insurance.	• People moved in with relatives and used shelters. • American people rebuilt their homes but used tradesmen rather than did it themselves. • They have home insurance but they also received aid from the government and other organisations.
Organisations	• UN provided US$5.5 million to Cuba from the CERF and US$1.5 million in emergency funds. • Venezuela sent 650 tonnes of aid, including non-perishable food, potable water and heavy machinery. • Cuban Red Cross delivered relief aid to approximately 25,000 families which included roofing materials, mattresses, clean drinking water, hygiene and kitchen kits. • World Food Programme immediately responded with US$1 million to assist the 788,000 people in the most affected areas of Cuba with a one-month food ration.	• Canadian Rivers Institute worked with a number of other NGOs and public agencies to restore beaches by clearing rubble and replenishing sand. • Red Cross had 17,000 trained workers, 90% of them volunteers who provided, amongst other aid, over 300 response vehicles, 74,000 overnight stays and 17 million meals and snacks. • AmeriCares, an American charity, responded quickly by sending teams of relief workers to hard-hit areas, sending aid shipments, providing funding and deploying a mobile medical clinic.
Governments	• Government sent teams of electricians to Santiago province from all over the island within hours of the hurricane hitting. • Government quickly made building materials available to residents, including corrugated iron sheets, metal rods and cement. • Local government officials compiled data from families about the damage they had experienced so that the government could send the appropriate help. • Military teams were mobilised quickly to clear the streets of rubble and an estimated 6.5 million cubic metres (230 million cubic feet) of felled trees.	• US government approved US$60.3 billion in aid to citizens affected by Hurricane Sandy. • FEMA teams and resources were put in place ready to help people even before the hurricane had caused any problems. They were on hand to offer any help that was needed. • FEMA and the Army Corps of Engineers worked to quickly reopen most of the beaches in New Jersey. • The Department of Agriculture promised US$6.2 million for emergency food assistance, infrastructure and economic programmes.

Figure 10 **Different responses of individuals, organisations and governments in Cuba and the USA to Hurricane Sandy**

CERF is the United Nations' Central Emergencies Response Fund.

Humanitarian aid is the help given after a natural disaster to save lives and reduce impacts.

FEMA is the USA'S Federal Emergency Management Agency.

The causes of drought are complex with some locations more vulnerable than others

What are the characteristics of arid environments compared to the extreme weather conditions associated with drought?

The characteristics of arid environments are:
- an average rainfall of between 100 and 300 mm
- variations in rainfall totals of between 50 and 100 per cent each year
- pastoral farming, usually by nomadic herdsmen
- natural vegetation is sparse – grasses, small shrubs and trees
- a short growing season of about 75 days.

The characteristics of areas experiencing drought are:
- a gradual reduction in the amount of available water supply
- the lack of water is unpredictable and to some extent unexpected.

> **Pastoral farming** is the farming of animals.
>
> **Nomadic herdsmen** are people who raise animals for their own food. They move around and have no fixed land.
>
> **Drought** is when a long period of unusually dry weather leads to a shortage of water.

What are the different causes of the weather hazard of drought?

Causes of drought	Definition
Meteorological	This concerns the amount of precipitation an area receives compared to its average. It is all about the weather and occurs if there is a prolonged period of below-average precipitation which creates a natural shortage of available water; this is then called a drought.
Hydrological	This is how a decrease in precipitation can have an impact on overland flow, reservoirs, lakes and ground water. This is often defined on a river basin scale: water reserves in aquifers, lakes and reservoirs fall below an established statistical average This can be related to precipitation or human demand and increased usage which has lowered water reserves.
Human – agricultural	This is when there is not enough water available to support average crop production on farms. This could be when the crops are planted or during their growing season; this often occurs when there is a fall in precipitation but can also occur if farming techniques change. For example, areas could start growing crops which require more water than is available but the water can be provided by irrigation. If the irrigation source dries up, then the plants will die.
Human – dam building	If a dam is constructed on a large river it can produce electricity and plenty of water for the area close to the dam. However, places further downstream may experience drought because they will be receiving a reduced flow of water. For example, the building of the Atatürk Dam on the River Euphrates provided electricity and water for irrigation in Turkey but has restricted the flow of water to Syria and Iraq, meaning they have less water for irrigation.
Human – deforestation	The cutting down of trees for fuel reduces the soil's ability to hold water. This can cause the land to dry out, which can result in drought in an area.

Figure 11 **Different causes of drought**

Irrigation is the artificial watering of the land.

Savannah ecosystem is an area of grassland which has a few shrubs and trees. It can be found in tropical areas.

Why does the global circulation make some locations more vulnerable to drought as a natural hazard than others?

REVISED

- The air rises at the Equator causing thunderstorms and a loss of moisture.
- The drier air moves north towards the mid latitudes.
- When it reaches approximately 30° N and 30°S, the dry air descends and warms.
- Many of the world's arid areas are found at these latitudes because of this air circulation.

The impacts of, and responses to, drought vary depending on a country's level of development

Why are droughts hazardous?

REVISED

There will be a shortage of water supplies and residents will be asked to conserve water. In extreme cases, stand pipes will be introduced.

During droughts, crops fail and animals die due to lack of grazing land. This can cause malnutrition and starvation.

Hazards of droughts

Wildfires are common. This is because the trees are very dry and burn easily. There are also a lot of fallen branches and dead wood lying in the forest. If a fire does start, there are only scarce water supplies to help with controlling the fire.

When soil becomes dry due to lack of rain, vegetation dies leaving the soil unprotected. The dry soil can be blown away by wind in a process known as wind erosion. When rain returns to the area, there is no top soil left so natural vegetation cannot regrow and crops cannot be planted.

Exam tip

Make short notes on the impacts and responses you are going to learn. Don't try to learn every fact, you only need to remember a few.

Impacts and responses to drought

Impacts	Namibia, drought of 2013, developing country	California, drought of 2014, developed country
People	• One in three people were at risk of malnutrition. • 778,000 Namibians were either severely or moderately food insecure. • Harvest yielded 42% less than in 2012, which meant severe food shortages.	• Loss of 17,100 jobs in farming. • 5% of the irrigated land in California won't be planted. • Fruit and vegetable prices will rise by 6%.
Environment	• Severe drought can have a great impact on a savannah ecosystem. It can change an area of grassland that could sustain livestock to an area of inedible grasses and plants which livestock cannot live on. This is because the grasses that can cope with drought are not good for livestock. • Large areas of Namibia are changing from savannah grasslands to desert due to the lack of rainfall. Only drought-resistant plants can survive in these conditions but the Namibian farmers' cattle cannot graze on them.	• Wildfires are becoming a more regular occurrence because of the dry and dead wood. • If water levels continue to drop in rivers, the water will become warmer and the young salmon will be unable to survive as they require cool running water.

Figure 12 **Impacts of drought in Namibia, 2013 and California, 2014**

	Namibia	California
Organisations	• In May, President Pohamba declared a state of emergency and requested US$1 million in international support to avert a crisis. • Unicef appealed for US$7 million to support their efforts to respond to the needs of women and children. • International Red Cross and Red Crescent Societies asked for US$1.5 million.	• New mandatory laws forbid restaurants to put water on tables without it being requested. • Hotels must also ask guest if they will reuse their linen and towels to save water. • They are also developing new ways to better manage and monitor the state's water resources.
Individuals	• Farmers have been forced to sell their livestock. • People have migrated to towns in search of work. • In one village almost all the people, about 350, left in search of water and grazing land for their cattle.	• Farmers will have to pump more water, which will cost an extra US$453. • People have been asked to use water more sparingly. • Farmers are planting fewer crops because there is not enough water for them to grow.
Governments	• Declared state of emergency and pledged £13 million in relief for the worst-hit households. • The Ministry of Agriculture, Water and Forestry (MAWF) gave farmers, who do not have enough grazing for their animals, two options: either to sell their livestock while they are still in good condition or financial help to transport them to other grazing areas.	• Governor Brown issued a state of emergency and in February, President Obama gave US$183 million from federal government funds. • In March, Governor Brown signed drought-relief legislation worth US$687 million. It included US$25.3 million for food.

Figure 13 **Responses to drought in Namibia, 2013 and California, 2014**

6 Ecosystems, biodiversity and management

Large-scale ecosystems are found in different parts of the world

What are the distributions and the characteristics
of the world's large-scale ecosystems?

REVISED

Ecosystem	Country	Climate	Fauna and flora
Tundra	Canada	• Temperature range between –34°C and 12°C • Total annual rainfall 200 mm, much of which falls as snow • Short growing season of about 60 days.	• Plant species which have shallow root systems and are low to the ground to cope with the harsh climate, such as mosses, lichens, grasses and dwarf shrubs • Animals which live here have adapted in different ways – brown bears eat in the summer and then store the food in a thick layer of insulating fat which they live off while they hibernate in the winter.
Boreal forests	Russia	• Temperature range between –10°C and 15°C • Total annual rainfall 500 mm.	• Evergreen trees, which allow growth to start when the weather warms up • Shallow root systems because of shallow soil and frozen ground • Trees such as pine and fir • Animals such as red foxes and black bears.
Temperate forests	USA	• Temperature range between 4°C and 17°C • Total annual rainfall 1,000 mm.	• Trees lose their leaves in winter to reduce transpiration • Vegetation is in four layers – canopy, sub-canopy, herb and ground. Many are dominated by one tree species – oak • Animals such as rabbits and deer.
Tropical forests	Brazil	• Temperature range between 27°C and 30°C • Total annual rainfall 2,200 mm.	• Vegetation is in four layers – emergents, canopy, under canopy and shrub/forest floor • Lianas wind their way up the trees • Epiphytes grow on the trees • Animals such as sloth and howler monkeys.
Temperate grasslands	Argentina	• Temperature range between 10°C and 18°C • Total annual rainfall 500 mm, most falling in the summer months.	• Trees are generally not found in these areas • Grasses such as purple needlegrass and buffalo grass grow in these areas • In North America, the temperate grasslands, known as the prairies, have been converted into farmland.

Figure 1 **Distributions and characteristics of the world's large-scale ecosystems**

Ecosystem	Country	Climate	Fauna and flora
Tropical grasslands	Kenya	• Temperature range between 25°C and 30°C • Total annual rainfall 1,000 mm • Rain is concentrated in six to eight months of the year; the rest of the year has drought conditions.	• Animals reproduce during the wet season when there is plentiful food and water. An example of this is the giraffe • Grasses grow very tall during the wet season, up to 2 m in height, but die off during the dry season • A few trees are found in these areas such as the Acacia tree which survives due to its thick trunk which holds water.
Deserts	Australia	• Temperature range between 30°C and 35°C • Great temperature differences between day and night, –18°C and 45°C • Very unpredictable rainfall, but generally very low.	• The only plants are short shrubs such as the prickly pear cactus which stores water in its spongy tissue • Animals which live in the desert, such as camels, store fat in their humps which they can change into water when it is needed.

Figure 1 **Distributions and characteristics of the world's large-scale ecosystems**

Hibernate is when an animal spends the winter in close quarters in a dormant (sleeping) condition.

Transpiration is evaporation of moisture from the leaf of a plant.

Ecosystem is a community of plants and animals and their non-living environment.

Altitude is height above sea level.

Distribution is where something is located.

Exam tips

Ensure that you know details about the world's ecosystems and examples of countries where you can find them.

How do climate and local factors, such as soils and altitude, influence the distribution of large-scale ecosystems?

REVISED

Climate – all ecosystems require different climates. The amount of rainfall and warmth an area receives will determine the type of plants that grow there.

Altitude – as land becomes higher, a number of changes take place in the abiotic factors:
● The temperature drops by 1°C per 100 m.
● Soils become thinner and contain less organic matter.
● The soil temperature also drops.

All of these have an impact on the vegetation that will grow.

Soils – underlying bedrock and soil are important for the type of ecosystem found in an area because they have an impact on the types of plants that will grow.

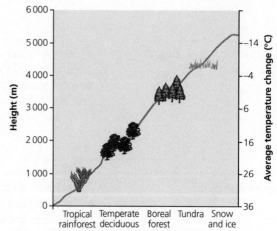

Figure 2 **Changes in vegetation type and temperature as altitude rises**

Exam practice

Explain how changes in altitude affect the distribution of ecosystems. Use Figure 2 in your answer.   (4 marks)

ONLINE

The biosphere is a vital system

How does the biosphere provide resources for people?

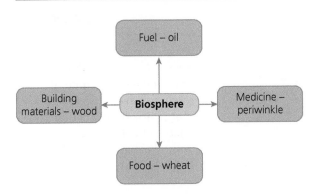

> **Biosphere** is the part of the Earth and its atmosphere in which living organisms exist or that is capable of supporting life.
>
> **Resource** is a stock or supply of something that is useful to people.

How is the biosphere being exploited commercially for energy, water and mineral resources?

Resource	Exploitation
Energy	• Oil is used for transportation and in the production of electricity. • Coal is used in the production of electricity. • Wind turbines have been built on land and sea to provide energy. • Solar panels have been put in fields to provide electricity using solar radiation. • More recently, the sea has been used to provide energy through the use of wave and tidal barrages.
Water	• Water is used for drinking, washing, toilets and cleaning. • It is used in the production of electricity in thermal power stations. • Water is also used by many industries, from the food industry to the paper industry. In Canada, the paper industry uses 45% of all water used by industry. • Farmers use water to irrigate their crops.
Mineral resources	Gold and silver are easily recognised as minerals used in the making of jewellery but silver is also in the making of mirrors. It is easy to see that the use of minerals is being exploited and certain minerals will soon start to be in short supply.

> **Exploitation** is the act of unfairly using resources for the benefit of people.
>
> **Mineral** is a solid, naturally occurring inorganic substance.
>
> **Finite resource** is one that will eventually run out.
>
> **Water cycle** is the closed system in which water moves between the atmosphere, the oceans and land.

Figure 3 **How the biosphere is being exploited for resources**

Now test yourself

TESTED

How is the biosphere being exploited?

> **Exam tip**
>
> Learn terms such as 'resource' and 'exploitation'.

The UK has its own variety of distinctive ecosystems

What are the distribution and characteristics of the UK's main terrestrial ecosystems?

Moorland

Found in highland areas with heavy rainfall such as:
Cairngorms in Scotland
North Yorkshire
Dartmoor
Mid Wales.

Soils are acid and peaty so only certain plants can survive, such as bell heather and bracken. Animals such as red deer and foxes. Birds such as buzzards and grouse.

Heathland

Found in lowland areas such as:
Cornwall
Devon
Cannock Chase in Staffordshire
The Gower in Wales.

Dry, sandy soil which is free draining, acidic and has few plant nutrients. Small shrubs such as heather and gorse and silver birch trees. Animals such as rabbits and hares. Birds such as nightjar and skylark. It is particularly important for reptiles.

Woodland

Native trees are broadleaved, deciduous trees such as oak and ash. And in Scotland, the Scots Pine, although broadleaved trees would also be found in lowland areas. There are also plantations of non-native conifers in many upland areas.

Trees are the dominant plant. Broadleaved trees tend to be deciduous. Mosses and lichens grow under the canopy and plants such as bluebells and ferns. Animals in the forest such as roe deer and badger. Birds such as sparrowhawk and tawny owl. Coniferous woods are made up of conifers which have needle-like leaves.

Wetland

Wetlands in the UK range from ponds and streams to rivers like the Severn and areas such as the Somerset Levels, East Anglian Fens and the Norfolk Broads.

Lowland fens have peaty, fertile soils which are periodically waterlogged. They support lush vegetation. Plants such as reeds and bulrush grow along the sides of the streams. Animals such as otters. Birds such as mallard and teal are common.

☐ = Distribution ☐ = Characteristics

Figure 4 Distribution and characteristics of the UK's main terrestrial ecosystems

Moorland is land which is not intensively farmed. It tends to have few trees; the plants are small shrubs such as heather. It is in upland areas of the UK and tends to have acidic, peaty soils.

Heathland tends to be open countryside in lowland areas. The plants are small shrubs, such as heather and gorse, with a few silver birch trees.

Deciduous woods are broadleaved trees, such as oak and ash, which lose their leaves in the autumn and regrow them each spring.

Coniferous woods are trees which stay in leaf all year round (evergreens). They tend to be confined to Scotland in their natural habitat but have been planted in other parts of the UK to be grown commercially.

Wetlands are areas of low-lying land which are predominantly wet and boggy. Some of the areas have been drained, such as the Somerset Levels and the Fens. This term also refers to small ponds and river estuaries.

How important a resource are marine ecosystems and how are human activities degrading them?

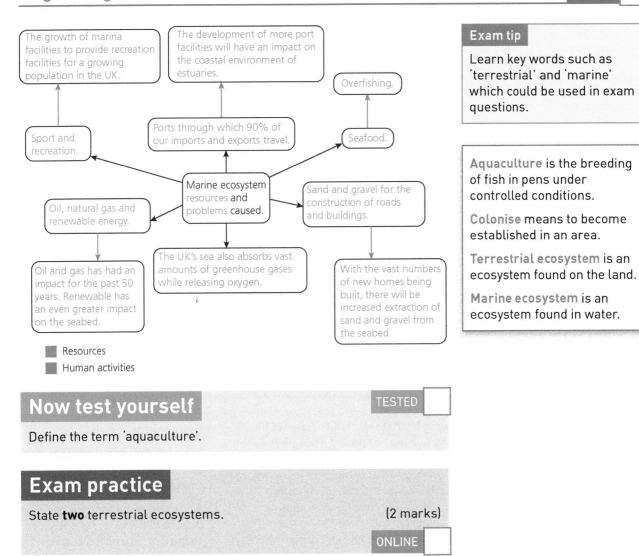

Resources
Human activities

Exam tip

Learn key words such as 'terrestrial' and 'marine' which could be used in exam questions.

Aquaculture is the breeding of fish in pens under controlled conditions.

Colonise means to become established in an area.

Terrestrial ecosystem is an ecosystem found on the land.

Marine ecosystem is an ecosystem found in water.

Now test yourself

TESTED ☐

Define the term 'aquaculture'.

Exam practice

State **two** terrestrial ecosystems. (2 marks)

ONLINE ☐

Distinguishing features of a tropical rainforest

What are the biotic and abiotic characteristics of the tropical rainforest ecosystem and how have plants and animals adapted to these conditions?

Abiotic characteristics	Biotic characteristics and adaptations
• Rains every day – total annual rainfall of 2,200 mm.	• The large trees have buttress roots which give them stability because of their great height. The roots are also a nutrient store. • The trees in the canopy have small leaves to prevent water loss through transpiration.
• Temperature range between 27°C and 30°C. • Very little light variation throughout the year – 12 hours' daylight, 12 hours' night.	• Plants on the forest floor have adapted. They have large leaves, due to lack of light and drip tips to help them to shed rainwater quickly.
• Soil poor quality; nutrients are washed down through the soil by the heavy rains. This forms a hard pan which is a layer of solid nutrient lower down in the soil which cannot be accessed by plants.	• The hummingbird lives in the canopy. It has strong flight muscles and does figure-of-eight wing beats to allow it to hover in the air. • Toucans live in the canopy. They have long bills/beaks to reach fruit on branches which are too small to support their weight. • The sloth lives in the canopy. It uses camouflage and amazing slowness to escape predators. They hang from branches in the canopy, and are so still that predators such as jaguars don't see them.

Abiotic factors are the physical, non-living environment such as water, wind, oxygen.

Biotic factors are the living organisms found in an area.

Organic material is something that was once living.

Inorganic material is something that was never living matter.

Soil is the top layer of the earth in which plants grow. It contains organic and inorganic material.

Litter is decomposing leaf and other organic debris found on forest floors.

Biomass is the amount or weight of living or recently living organisms in an area.

Figure 5 **Biotic and abiotic characteristics of the tropical rainforest ecosystem and adaptations of plants and animals**

Revision activity

Draw a table with three columns: 'Abiotic', 'Biotic' and 'Adaptations'. Put in information that you are going to learn. Remember, you will be given a lot of information, but you don't need to learn all of it! Choose two animal and two plant adaptations that you think you will be able to remember and concentrate on them.

Exam practice

Define the term 'biotic factors'. (1 mark)

ONLINE

What are the characteristics of the tropical rainforest food web?

REVISED

A food chain is the sequence of who eats who in an ecosystem to obtain the energy to survive. A network of food chains is known as a food web. The food web starts with plants which are known as producers; they gain their energy from the Sun. Plants are eaten by herbivores or primary consumers. Primary consumers are eaten by secondary consumers, which in turn may be eaten by tertiary consumers. When an organism dies, it is eaten by tiny microbes which are known as detrivores.

> **Nutrient cycle** is the movement and exchange of material, both organic and inorganic, into living matter.
>
> **Food chain** is a series of steps by which energy is obtained and used by living organisms.
>
> **Food web** is a network of food chains by which energy and nutrients are passed from one species to another; it is essentially who eats who.
>
> **Limiting factors** are factors such as temperature, moisture, light and nutrients. All of these are in abundance in tropical rainforests, which accounts for their high biodiversity.

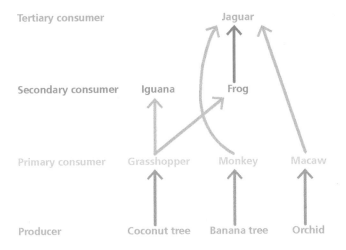

Figure 6 A tropical rainforest food web

Now test yourself

TESTED

State an animal or plant in each stage of the food web of the tropical rainforest.

What is the nutrient cycle?

REVISED

Nutrients are chemical elements and compounds which are needed for organisms to grow and live. The nutrient cycle is the movement of these compounds from the non-living environment to the living environment and back again.

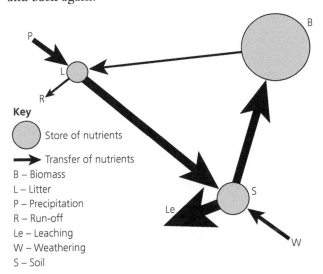

Key

◯ Store of nutrients
➤ Transfer of nutrients
B – Biomass
L – Litter
P – Precipitation
R – Run-off
Le – Leaching
W – Weathering
S – Soil

Figure 7 The nutrient cycle in the tropical rainforest

Goods and services are under threat in tropical rainforest ecosystems

Structure of the tropical rainforest is the layers of plants and animals in the forest.

Function of the tropical rainforest is its ecosystem and how it works.

Biodiversity is the number of species present in an area.

Glaucoma is a condition which gradually causes someone to lose their sight.

Eutrophication is the growth of algae on water courses due to the increase in chemical fertilisers being used on the land.

Monoculture is the growing of one crop on large areas of land.

Overpopulation is too many people living in an area for the area to support.

Favelas are Brazilian houses made of waste materials, also known as shanty towns.

Which goods and services are provided by tropical rainforest ecosystems?

REVISED

The rosy periwinkle which is found in Madagascar has properties which can halt the progress of Hodgkin's disease in 58% of cases. Sales of medicines from this one plant are about US$160 million a year.

Medicines
120 prescription medicines sold globally are derived directly from rainforest plants.

Quinine which helps to cure malaria is an alkaloid extracted from the bark of the cinchona tree found in tropical forests in South America and Africa.

Bananas grow in the tropical forest. They have become increasingly popular. They are a US$5 billion global industry.

Food stuffs

Palm oil which is grown in many tropical rainforest regions of the world is used globally in food products such as pizza dough, biscuits and bread.

Wood from trees such as mahogany and teak is used for flooring and furniture in temperate regions, such as the USA and the UK.

Timber

Local people use wood from the rainforest for building materials and fuel. This is becoming more of a problem as population numbers in the tropical areas continue to rise.

Zip wires are constructed through the canopy and under the canopy for tourists to enjoy. There are also hanging bridges and courses for them to walk around.

Recreation
The rainforest provides many opportunities for recreational activities. A number of countries have developed the rainforest in this way to provide an income which does not involve deforestation.

Nature trails on the forest floor or in the canopy are organised with trained guides who teach the tourists about the rainforest.

How does climate change present a threat to the structure, functioning and biodiversity of tropical rainforest ecosystems?

REVISED

Climate change presents a threat because the structure and function of the tropical rainforest rely on the climate that the area receives. Therefore, if there are changes in temperature and rainfall distribution, the rainforest will be unable to survive in its present form. The rainforest is vulnerable to climate change because its resilience has been weakened by human activities such as deforestation. If part of the forest is felled or is damaged by a fire, it has an impact on the rest of the forest. The humidity in the rainforest and the vast amount of transpiration means that much of the rainfall is recycled from the forest itself. If the forest is felled, there are fewer trees to provide water and therefore less rainfall occurs. Climate change could see eastern areas of the Amazon rainforest receiving 20 per cent less rainfall by 2030. This will cause the temperature to rise and will have a major impact on the forest in that area. The Amazon rainforest contains 40 per cent of the species on earth. This biodiversity will be threatened by less rainfall and higher temperatures because links in the food chain could be broken if species fail to adapt to the new climate conditions. The predictions are that deforestation and climate change could damage or destroy 60 per cent of the Amazon rainforest by 2030.

Now test yourself and exam practice answers at **www.hoddereducation.co.uk/myrevisionnotes**

What are the economic and social causes of deforestation?

Economic causes of deforestation are related to the country using its natural resources to generate income.	Social causes of deforestation relate to population pressure from the growing population of the world.
Tropical rainforest is cleared for agriculture. The forest is felled and burned; this adds nutrients to the soil which last for a few years. The cleared area is quickly planted and good crops are harvested for a few years. After this time, large amounts of fertiliser are needed to produce high yields. This makes the farming less profitable and causes fertilisers to be washed into streams and rivers, causing eutrophication. The land is then abandoned or left to cattle ranchers.	The main reason for deforestation is population pressure due to overpopulation. This relates to the growth in population in the temperate areas of the world such as the UK, which demand rainforest products, and the population of the country where the tropical rainforest is situated, which grows and makes more demands on the forest area. It is predicted that the world's population will reach 8 billion by 2026. In Brazil, the biggest increases in population are in the Amazon. Ten cities in the Brazilian Amazon doubled their population between 2000 and 2010.
Resources have been extracted from rainforest areas for many years. The governments of countries such as Brazil have been paid for the rights to the minerals. The governments have seen it as a way of raising money to help develop their country; however, the people who live in the rainforest have not been consulted. This has caused many problems because the indigenous inhabitants do not believe that the government owns the land and has no right to grant mineral rights to large companies. The large-scale mining operations have resulted in deforestation through the clearing of the forest for the mines but also for the roads and settlements for the workers. The miners have also brought problems such as diseases to the local inhabitants.	

Figure 8 **Economic and social causes of deforestation**

The sustainable management of the tropical rainforest: Costa Rica

Direct government action – tax deductions and grants to owners of rainforest if they conserved their forest area and used it to benefit society. The government issues landowners forest protection certificates and pays them US$50 annually for every hectare of forest they protect.

There are 11 eco-regions in Costa Rica created by the government. Each area can make decisions on how its rainforest will be protected.

The Certificate for Sustainable Tourism (CST) for businesses proves their commitment to sustainable tourism.

Sustainable management of the tropical rainforest in Costa Rica

The government set up national parks which protect 18 per cent of the country and privately-owned reserves which protect another 13 per cent.

Another way that Costa Rica is using its rainforest as a commodity is by selling carbon credits. Wealthy countries buy them to offset the carbon that they produce. This is a way for Costa Rica to earn money from its rainforest without cutting down the trees. In 1999, this idea generated US$20 million for the country.

Deforested areas of the rainforest are being used to economically support the country; for example, for farming commercial crops such as bananas.

Ecotourism is travel to natural areas which does no damage, it conserves the environment and improves the wellbeing of local people.

Carbon credits is a permit which allows the holder to emit one ton of carbon dioxide or another greenhouse gas. Carbon credits can be traded between countries or businesses.

For example, a steel producer in the USA has been allowed to emit 10 tons of carbon dioxide but knows it will emit 11 tons. The company could buy one credit from Costa Rica to ensure that it keeps to international law. Costa Rica has many carbon credits because of its rainforest. In this way, wealthy countries are encouraging poorer countries to protect their rainforest.

NGO is an organisation that is not part of the government or a profit-making business. It is usually set up by private individuals and can be funded by donations or governments.

Exam tip

Remember, don't try to learn all the information for a case study. Choose the information, which may be facts or figures or a mixture of both, and learn that. It would be helpful to draw up your own table of information or underline the information on this page that you are going to learn.

Distinguishing features of deciduous woodlands

What are the biotic and abiotic characteristics of the deciduous woodland ecosystem and how have plants and animals adapted to these conditions?

REVISED

Abiotic characteristics	Biotic characteristics and adaptations
• Total annual rainfall 1,000 mm.	• In spring, deciduous trees grow thin, broad, light-weight leaves. These leaves capture the sunlight easily and allow the tree to grow quickly as the temperature warms and the days grow longer. However, these leaves have too much exposed surface area for the cold winter months and therefore the tree loses its leaves as the weather becomes colder and the daylight hours shorter. • Sub-canopy layer – trees such as rowans and dogwoods and shrubs such as rhododendrons.
• Temperature range between 4°C and 17°C. Long periods of light in the summer, approximately 18 hours, contrasting with short days in the winter of about eight hours of light.	• Field or herb layer – plants in this layer flower early in the year before the trees in the canopy have grown their leaves, which block out the light.
• The soil is fertile. The autumn leaf fall means that there are plenty of nutrients. • Earthworms in the soil help to mix the nutrients. • The tree roots are deep and therefore help to break up the rock below, which gives the soil more nutrients.	• Hedgehogs hibernate during the cold winter months from about November to April. • The nightingale migrates to countries in Africa in September, returning to the UK in April to avoid the cold months when the woodlands offer little food. • Squirrels store food in the ground under fallen leaves so that they have food in the colder months.

Figure 9 Biotic and abiotic characteristics of the deciduous woodland ecosystem and adaptations of plants and animals

Revision activity

Draw a table with three columns: 'Abiotic', 'Biotic' and 'Adaptations'. Put in information that you are going to learn. Remember, you will be given a lot of information, but you don't need to learn all of it! Choose two animal and two plant adaptations that you think you will be able to remember and concentrate on them.

Exam practice

Explain **two** ways in which plants have adapted to living in deciduous woodlands.　　(4 marks)

ONLINE

What are the characteristics of the deciduous woodland food web?

A food chain is the sequence of who eats who in an ecosystem to obtain the energy to survive. A network of food chains is known as a food web. The food web starts with plants which are known as producers; they gain their energy from the Sun.

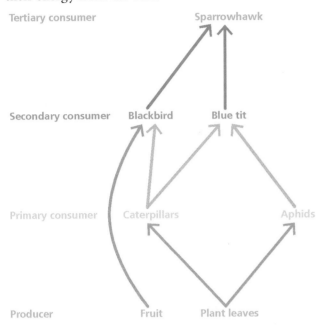

Figure 10 **A deciduous woodland food web**

Now test yourself

State an animal or plant in each stage of the food web of the deciduous woodland.

TESTED

What is the nutrient cycle?

Nutrients are chemical elements and compounds which are needed for organisms to grow and live. The nutrient cycle is the movement of these compounds from the non-living environment to the living environment and back again. In the deciduous woodland, nutrients are stored in the biomass and soil in almost equal amounts with a slightly smaller store in the litter.

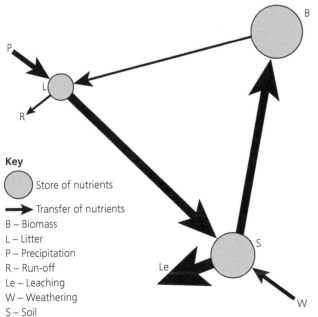

Key

◯ Store of nutrients

➤ Transfer of nutrients

B – Biomass
L – Litter
P – Precipitation
R – Run-off
Le – Leaching
W – Weathering
S – Soil

Figure 11 **The nutrient cycle in a deciduous woodland**

Goods and services are under threat in deciduous woodland ecosystems

Which goods and services are provided by deciduous woodland ecosystems?

Woodlands in the UK are managed as a resource for the goods and services that they can provide. They are also managed for the benefit of the wildlife that lives within them.

In 2009, a total of 0.4 million tonnes of timber was produced from deciduous woodlands in the UK. ←→ **Timber** → Timber is increasingly used in the construction industry to build houses.

Short rotation coppice willow was developed in the 1990s to provide biomass for domestic heating. 7 to 9 tonnes of wood a year are needed to heat an average three-bedroom house. A small coppice can produce three tonnes a year. This is popular in Northern Ireland. ←→ **Fuel** → Of the 0.4 million tonnes of broad-leaved timber cut down for commercial use in the UK in 2009, 69% was used as fuel.

In the UK, over half the area of woodland has public access. This is carefully managed to ensure that the woodlands are conserved and managed correctly. ←→ **Conservation and recreation** → Recreational visits to woodlands were valued at £484 million in 2010.

Broad-leaved trees are deciduous trees which lose their leaves in winter, such as oak and elm.

Short rotation coppice is usually willow trees which are grown specifically to be turned into wood chip for biomass boilers for domestic heating or power stations. They are planted densely and harvests are on two- and five-year cycles.

Ancient woodlands contain trees which were planted before 1600.

Exam practice

State **two** goods or services provided by temperate deciduous woodlands. (2 marks)

Exam tip

There is a lot of information here and more in your textbook. Choose the uses you want to learn. Include one use from each of the three topics – timber, fuel and conservation and recreation.

How does climate change present a threat to the structure, functioning and biodiversity of deciduous woodland ecosystems?

REVISED

Recent climate change has not had a major impact on woodland structure and functioning although small changes can be identified. The lack of information about the impact of climate change on deciduous woodlands is partly due to the fact that the trees are long-lived and can adapt to climate variability. However:

- Increasing temperatures have led to faster tree growth in some areas, but overall there has been little impact.
- More storms have occurred, which can have an impact on the stability of the trees.
- Increased droughts in the summer have had a detrimental effect on tree growth in some areas. It is also feared that the stress that droughts put on trees may make them more susceptible to disease.

- Milder winters could cause problems as many trees need cold weather to help them to reset their clocks for spring. Without this, fruiting and flowering may be disrupted. Milder winters also means that pests and diseases are not killed in winter frosts.
- It is predicted that during the twenty-first century temperatures in the UK will rise by 2.5°C. This will mean that species will have to move north by 300 km or 300 m uphill to attain the growing conditions they require or adapt to the new conditions.

What are the economic and social causes of deforestation?

REVISED

Economic causes	Social causes
The farming landscape of the UK has changed many times since the Middle Ages. As farming has changed though the centuries, the amount of forested area in the UK has declined. Some parts of the UK have kept more of their forested area, but overall the forested area declined to a low of 5% in 1919.	As the population grew in the Middle Ages there was a greater demand for housing. This meant that trees had to be felled to provide beams to support the roofs. Forests were also cleared to make way for towns, especially in the north of England where the Industrial Revolution took place. Population growth between 1945 and 1975 meant that many of the remaining deciduous woods were cut down to make way for suburbs in the existing towns and cities, or for the 'new towns' that were built during that period. For example, Bracknell, a new town, was built by clearing large areas of Windsor Forest. Other cities such as London (2,500 hectares) and Sheffield (650 hectares) have retained some in parks and other woodlands, which are now used for recreational purposes.
Timber has been extracted from forests of the UK for centuries. It was first used for house building and fuel. English oaks were also used in shipbuilding and the rise of the British Empire saw a great demand for timber to build ships in the late sixteenth century. A survey of the New Forest in 1608 found almost 124,000 trees fit for navy timber. By 1707 that figure had declined to less than 12,500.	
Further timber extraction occurred during the First World War as it was needed to build trenches. This led to the forested area of the UK being at an all-time low at the end of the war. Since this time, the Forestry Commission has planted millions of hectares of land with fast-growing conifers. Since the 1970s the Forestry Commission has changed its planting policy and as trees are extracted for timber, they are replaced with broad-leaved trees such as oak and elm in an attempt to re-establish deciduous woodlands in the UK.	

Figure 12 **Economic and social causes of deforestation**

Now test yourself

TESTED

How and why are deciduous woodlands being cut down?

The sustainable use and management of deciduous woodlands: Wyre Forest, West Midlands

The Wyre Forest is the largest area of ancient woodlands in England. It covers an area of 2,400 hectares. There are also stands of conifer plantations within the forest boundary and areas of orchard, meadows and mixed farming, making a total land area of nearly 5,000 hectares. The Forest is situated in the West Midlands to the west of the Birmingham conurbation and lies on the borders of Worcestershire, Shropshire and Staffordshire. The Forest is bisected by the River Severn with Bewdley on its eastern side and Cleobury Mortimer on its west.

Wildlife management

- Wildlife-rich meadows and orchards will be extended.
- Invasive species such as Himalayan balsam will be removed.
- A network of woodland rides will provide corridors for wildlife as well as people.

Community management

- The people who live in the area will be encouraged to take part in conservation work.
- There will be community woods in which local people can cut their own firewood.

Management of the woodland is by the Wyre Forest Landscape Partnership (WFLP).

Woodland management

- On steep slopes deciduous woodland will be left unmanaged to develop undisturbed.
- Areas previously planted with conifers will be gradually restored to woodland with oak as the predominant tree canopy.
- Other trees such as silver birch, aspen and rowan will be encouraged with an understorey of hawthorn, hazel and holly.
- The deer population and the non-native invasive plant population will be carefully controlled.

Leisure and recreation management

- The forest will provide a place where people of all abilities can go for leisure and recreation.
- The visitor centre at Callow Hill will have displays to help people understand the forest and what it has to offer them.
- The Forestry Commission provides a number of recreational activities including walking trails, cycle paths and a play area. The forest now has a GoApe experience with zip wires, Tarzan swings and walkways.

Education

- Monitoring takes place on the impact of using woodlands for recreation.
- Children and adults from the surrounding communities, especially Birmingham, have been introduced to woodlands and wildlife through interactive displays and workshops.
- It provides many opportunities for skills development and training including forest industries apprentices and internships.

Exam tip

Remember, don't try to learn all the information for a case study. Choose the information, which may be facts or figures or a mixture of both, and learn that. It would be helpful to draw up your own table of information or underline the information on this page that you are going to learn.

7 Changing cities

Urbanisation is a global process

What were the trends in urbanisation over the last 50 years in different parts of the world?

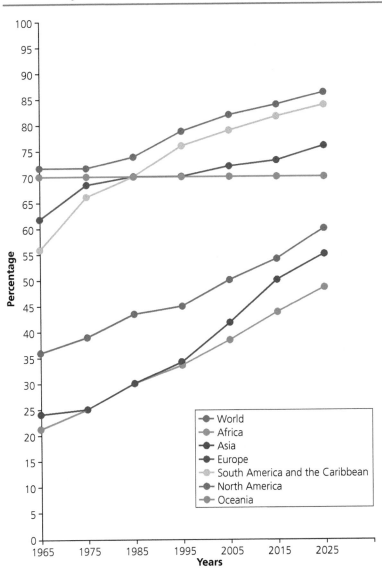

Figure 1 **Percentage of urban population per continent**

Developed country is a country with very high human development (VHHD).

Natural increase is when there is a positive difference between the birth rate and the death rate.

Urbanisation is the process by which increasing numbers of people within a country live in cities.

Developing country is a country with low human development (LHD), a poor country.

Emerging country is a country with high and medium human development (HMHD).

Major city is a city with a population of at least 400,000.

HDI (Human Development Index) is a measurement of how much progress a country has made.

Rural depopulation is the movement of people from rural to urban areas.

Now test yourself

Study the graph in Figure 1.
1 Which **three** continents had the highest urban populations in 1965?
2 Which **three** continents are experiencing the fastest growth of urban population?

Exam practice

Study Figure 1. Which continent has the fastest growing urban population?   (1 mark)

How and why has urbanisation occurred at different times and rates in different parts of the world?

REVISED ☐

Urbanisation in developed countries

- Occurred during the nineteenth century.
- Caused in part by the Industrial Revolution and the huge demand for labour in the new factories.
- Landless, poor villagers moved in great numbers to the cities.
- In the past 50 years, developed countries have continued to increase their urban areas but at a much slower rate.

Urbanisation in emerging and developing countries

- Occurred over the past 50 years.
- Main reason for this growth in urban population is the increase in population.
- Population growth has occurred because of decreasing death rates as more children who are born survive past their first birthday.
- There has also been a large natural increase in the population of urban areas, partly due to the fact that many of the people who live there are of child-bearing age.

Effects of high rates of urbanisation

Developed countries

Overcrowded cities – cities in many developed world countries are not coping with the vast amounts of people who wish to live there.

Housing – house prices and also rents of flats are increasing all the time. This is because there are not enough housing units for all the people who want them.

Transport – buses and tubes are not able to cope; people waiting on the platforms watch the tube trains going past full.

Education – schools do not have enough places for all of the children that need them. There are long waiting lists for the best schools in an area.

Developing and emerging countries

Agriculture – older people are left in the countryside because the young have gone to the towns looking for work. Soon food supplies will drop because the people who are left cannot work on the land.

Shanty towns – people who move to the cities cannot afford to rent homes and there is a lack of housing so they build their own homes on wasteland from waste materials.

Unemployment – in Cairo there are so few jobs that people make a living by picking over the rubbish in waste heaps to find things that they can sell.

Education – there are not enough places in the schools so in some countries only the boys go to school.

Exam tip

Remember, you will be given a lot of facts about the effects of urbanisation, but you don't need to learn all of them! Learn a couple of facts for developed countries and a couple for developing countries.

Now test yourself

TESTED ☐

What are the effects of high rates of urbanisation in developing countries?

The degree of urbanisation varies across the UK

How is the urban population of the UK distributed and where are its major urban centres?

REVISED

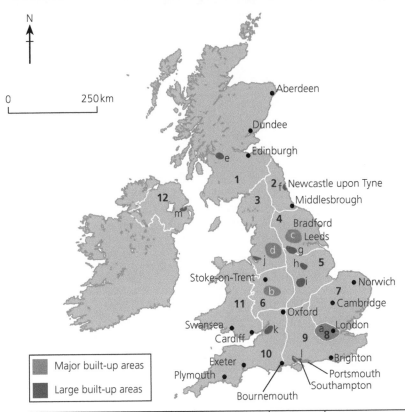

Figure 2 **The UK's major urban (built-up) centres**

Built up areas	Letter on map	Region of the UK	Number on map	Number of residents in 2011
Greater London	a	London	8	9,800,000
West Midlands	b	West Midlands	6	2,450,000
West Yorkshire	c	Yorkshire and Humber	4	1,800,000
Greater Manchester	d	North West	3	2,550,000
Glasgow	e	Scotland	1	1,200,000
Tyneside	f	North East	2	770,000
Sheffield	g	North West	3	680,000
Nottingham	h	East Midlands	5	720,000
Leicester	i	East Midlands	5	508,000
Liverpool	j	North West	3	860,000
Bristol	k	South West	10	600,000
South Hampshire	l	South East	9	850,000
Belfast	m	Northern Ireland	12	580,000

Distribution is the arrangement of something.

Density of population is the number of people in an area, usually given as people per square kilometre.

Region is a unit within a country.

Major urban centre is an area which has a high density of population and is made up of houses, industrial buildings, factories and transport routes. These areas are now referred to as built-up areas.

Rate of urbanisation is the speed at which settlements are built.

Degree of urbanisation is the amount of built-up area that has developed in a region.

Enclosure Acts were a series of Acts of Parliament between 1750 and 1860 which stopped villages using the open fields and commons that they had been allowed to use for centuries. This meant that they could not make a living and had to move to find a better life. Many of the villagers went to live in industrial towns.

Revision activity

Study Figure 2. Try to remember where the major urban centres of the UK are. One way of doing this is to draw a triangle to represent the UK and mark on it where the major urban centres are.

What factors cause the rate and degree of urbanisation to differ between the regions of the UK?

City	Reason for growth in the eighteenth and nineteenth centuries	Reason for growth in the twentieth century
Manchester Bradford	Industrial Revolution	
London Bristol Newcastle upon Tyne	Port development	
Swansea Cardiff	Natural resources – coal and iron ore	
Aberdeen		North Sea oil deposits
Blackpool Brighton Scarborough		Tourist industry
Eastbourne Bournemouth		Retirement settlement

Figure 3 **Factors which caused urbanisation to differ between the regions of the UK**

Exam practice

Explain the factors which caused urbanisation to differ between the regions of the UK. (4 marks)

ONLINE

8 Case study of a major UK city: Bristol

The context of the city of Bristol influences its functions and structure

What is the national, regional and global context of Bristol?

- National context – Bristol is in the southwest of the country, southeast of the Severn Estuary.
- Regional context – Bristol is in the United Kingdom in the northwest of Europe. It is to the north of France and west of Denmark.
- Global context – Bristol lies east of Canada in North America and west of Russia in Asia.

> **Site** is the land that the settlement is built upon.
>
> **Situation** is where the settlement is compared to physical and human features around it.
>
> **Connectivity** is the way that Bristol is connected or linked to other settlements in the UK and to other countries in the world.
>
> **Public buildings** are buildings owned by the council that serve the residents of the city, such as a library.
>
> **Residential** is when an area is used for housing.

Site, situation and location of Bristol

Bristol developed as a trading settlement with Spain, Portugal and colonies in the New World. The city still has a large port and docks at Avonmouth and Portbury. Bristol has excellent railway links with the rest of the UK with two major stations – Bristol Temple Meads and Bristol Parkway. London, Scotland, Wales, Manchester, Birmingham and Exeter are easily accessible by train.

The original settlement of Bristol grew up on the confluence of the River Avon and the River Frome. There are seven hills which are formed from the valleys of the two rivers and their tributaries. The limestone ridge to the west of the city is the most well-known hill. This is because the River Avon cuts through this forming the Avon Gorge over which is the Clifton Suspension Bridge.

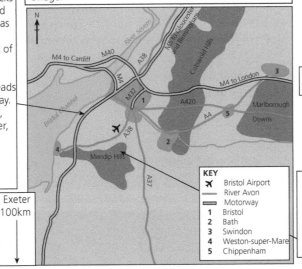

London 150km
Swindon 50km

Exeter 100km

Planes fly to 112 countries from the international airport which is to the southwest of the city.

KEY
- ✈ Bristol Airport
- — River Avon
- ▬ Motorway
- 1 Bristol
- 2 Bath
- 3 Swindon
- 4 Weston-super-Mare
- 5 Chippenham

Figure 1 **The site, situation and location of Bristol**

Now test yourself

What is the difference between the site and situation of a settlement?

What are the functions and building age of different parts of Bristol's structure?

Area of city	Functions	Building age
Central Business District (CBD)	● Shops ● Offices ● Financial institutions ● Entertainment facilities ● Public buildings which include the city museum and the council house.	● Some buildings date back to the original settlement but much of the centre of Bristol was rebuilt after the Second World War because the city centre was heavily bombed. ● Broadmead shopping centre was built in the 1950s. ● The Galleries shopping centre opened in 1991 but was redeveloped in 2013. ● Cabot Circus shopping centre opened in 2008.
Inner city	● Residential ● Small light industry.	● High-density housing built between 1850 and 1914.
Inner suburbs	● Residential ● Open spaces for parks, playing fields ● Schools ● Hospitals.	● Built between 1920 and 1940.
Outer suburbs	● Residential.	● Built from the 1960s onwards.
Urban-rural fringe	● Green belt land ● New housing estates ● Out-of-town shopping areas ● Transport routes ● Industrial estates.	● Built from the 1960s onwards. ● Built from the 1990s onwards.

Figure 2 **Functions and building age of different parts of Bristol's structure**

Terraced houses are houses which are joined on each side to the house next to them; their front doors usually open straight onto the street.

Semi-detached houses are houses that are joined on just one side to another house.

Detached houses are houses that are not joined to another house.

Owner occupied is when people own the house they live in or are buying it with a mortgage.

Social priority housing is houses that are owned by a housing association and rented to people who cannot afford to buy their own home.

Green belt land is an area around the city, which is composed of farmland and recreational land. There are strict controls on the development of this land. Its purpose is to control the growth of cities.

> **Exam tip**
>
> Remember, don't try to learn all the functions! Just learn a couple for each of the city areas.

Exam practice

Compare the functions of the CBD with that of the rural-urban fringe.

(3 marks)

Bristol is being changed by movements of people, employment and services

What is meant by the terms urbanisation, suburbanisation, counter-urbanisation and re-urbanisation and how have these processes had an impact on Bristol?

- **Urbanisation** – the increase in the number of people living in towns and cities compared to the number of people living in the countryside.
- **Suburbanisation** – the growth of a town or city into the surrounding countryside, which usually joins it to villages on its outskirts, making one large built-up area.
- **Counter-urbanisation** – the movement of people from cities to countryside areas.
- **Re-urbanisation** – the movement of people back into urban areas, usually after the city has been modernised.
- **National migration** – the movement of people from one area of a country to another with the intention of staying there for at least a year.
- **International migration** – the movement of people from one country to another with the intention of staying there for at least a year.

Year	Population	Year	Population
1086	1,500	1931	384,200
1400	9,600	1951	422,400
1600	15,000	1971	428,000
1801	64,000	1991	396,500
1901	323,698	2011	428,100

Figure 3 **Bristol's population growth, 1086–2011**

Slave triangle describes a three-part journey: ships left British ports such as Bristol with goods such as cloth and guns, they sailed to countries in Africa where these goods were sold and enslaved people were bought. The people were taken to the Caribbean and sold. Goods such as sugar were then bought with the money and brought back to England.

Emigration is when someone leaves a country.

Immigration is when someone moves into a country.

Median age range is the middle age range, if all ages are put in a line.

Now test yourself

Study Figure 3. In which year is there evidence of the following processes?
- Urbanisation
- Counter-urbanisation
- Re-urbanisation

Exam practice

Define the term 'counter-urbanisation'. (1 mark)

ONLINE

Exam tip

All of these key terms are important, so make sure that you learn them.

What are the causes of national and international migration?

Push factors	Pull factors
Natural hazards – flooding, drought	Hazard-free areas of the world
War and political conflicts – lack of safety	Political stability
Lack of jobs	Better job opportunities
Lack of facilities – education, medical, housing	Plenty of facilities
High crime rates	To be closer to family
Poverty	Good climate
Crop failure	Fertile land
Pollution	More attractive quality of life

Figure 4 **Causes of migration**

Exam practice

Explain **one** push factor and **one** pull factor of migration.

(4 marks)

What is the impact of national and international migration on different parts of Bristol?

- The population of Bristol has risen by 38,000 since 2001.
- The number of people living in Bristol who were born outside of the UK has risen by over eight per cent since 2001 bringing the total to 15 per cent of the population, of which 61 per cent arrived in the UK in the last ten years. This puts pressure on schools and other welfare provision because of issues such as English being the second language.
- Of this number, 69 per cent arrived in the UK when they were of working age, therefore more jobs were required to cope with this influx of people, and 30 per cent arrived as children under the age of 16. Therefore, school intake rose dramatically.
- Since 2008 there has been an increase in internal migration with people moving to Bristol from other parts of the country.
- The inner–city wards of Cabot and Lawrence Hill have seen a large increase in population from people who were born outside the UK. Almost 33 per cent of the increase in population in Bristol has taken place in these two wards. This has caused a strain on the housing stock that is available.

Now test yourself

Study the impacts of migration on Bristol. Explain the impacts on different areas of the city.

Globalisation and economic change create challenges for Bristol that require long-term solutions

Deindustrialisation is the reduction of industrial activity in a region.

Globalisation is the way that companies, ideas and lifestyles are being spread around the world.

Decentralisation is the process of spreading or dispersing power or people away from the central authority.

Edge- and out-of-town shopping are when shops or facilities are located away from the city centre, on the edge of cities.

Internet shopping is a way of shopping from home which allows people to buy or sell goods using the internet.

Energy-efficient housing is when less energy is used to provide the same level of heat or power for cooking. Homes are well insulated.

Bristol's key population characteristics

Ages	Males	Females
0–5	41,600	40,200
16–24	33,500	33,900
25–49	85,600	80,500
50–64	32,000	32,200
Over 65	25,800	32,200

Figure 5 **Bristol's age and gender structure, 2013**

Key characteristics of the population are:
- One in every five people living in Bristol is under the age of 16.
- The number of children under the age of 16 is more than the number of people over 65.
- Bristol has a higher percentage of people in the working age range than the rest of the UK.
- People aged 20 to 39 make up 36 per cent of the population. The average for rest of the UK is 29 per cent.

Why has Bristol's population grown?

- International migration caused the growth during the 2000s.
- An increase in the birth rate and a decrease in the death rate in the city is causing the growth now.

What are the causes of deindustrialisation and their impacts on Bristol?

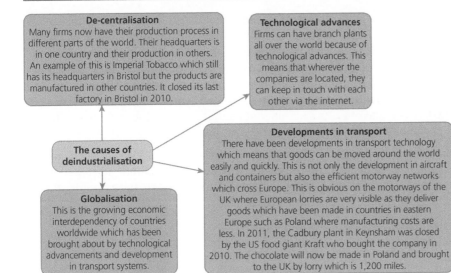

De-centralisation
Many firms now have their production process in different parts of the world. Their headquarters is in one country and their production in others. An example of this is Imperial Tobacco which still has its headquarters in Bristol but the products are manufactured in other countries. It closed its last factory in Bristol in 2010.

Technological advances
Firms can have branch plants all over the world because of technological advances. This means that wherever the companies are located, they can keep in touch with each other via the internet.

The causes of deindustrialisation

Globalisation
This is the growing economic interdependency of countries worldwide which has been brought about by technological advancements and development in transport systems.

Developments in transport
There have been developments in transport technology which means that goods can be moved around the world easily and quickly. This is not only the development in aircraft and containers but also the efficient motorway networks which cross Europe. This is obvious on the motorways of the UK where European lorries are very visible as they deliver goods which have been made in countries in eastern Europe such as Poland where manufacturing costs are less. In 2011 the Cadbury plant in Keynsham was closed by the US food giant Kraft who bought the company in 2010. The chocolate will now be made in Poland and brought to the UK by lorry which is 1,200 miles.

Exam practice

Study Bristol's population dynamics. Why has death rate fallen so dramatically?   (2 marks)

ONLINE

How is economic change increasing inequality in the city and causing differences in the quality of life for its residents?

The move from an industrial–based city with a large percentage of the population working in the secondary sector to a city with a large percentage of the population employed in the tertiary and quaternary sectors has caused an increase in inequality in the city. This is because many people who worked in industries which have now closed have not been able to find new employment because they lack the skills. This has caused inequality and a decline in the quality of life for this segment of the population.

What are the recent changes in retailing and their impact on Bristol?

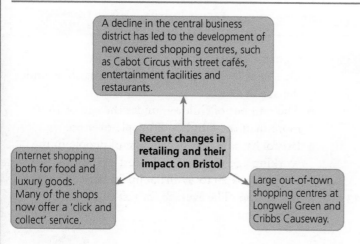

What is the range of possible strategies aimed at making urban living more sustainable and improving quality of life for the residents of Bristol?

Factor	Strategy
Employment	Bristol's unemployment rate is 8% which is one of the lowest rates in the country. This is due partly to the work of the council, which has been active in attracting companies to the city. The council has promised that Bristol will be the most sustainable city in the UK by 2020. This has attracted many green companies to the city. Bristol also has the highest growth of disposable income in the UK, with an average salary of £22,293 which is above the UK average of £21,473. This means that the population who live in Bristol have a high disposable income which generates jobs in service industries.
Education and health	As part of the council's 2020 vision, they promised to improve education standards in the city's schools and to provide better health care. The council has funded projects in schools about healthy eating. This has improved people's awareness, for example, the importance of eating five fruit and vegetables a day.

Figure 6 **Strategies to make urban living more sustainable and improve quality of life for the residents of Bristol**

Factor	Strategy
Recycling	Bristol council provides different bins for waste in the city and has done for many years. It also has a good provision of recycling centres spread around the city for people to take their waste to. The council also provides a collection service for large bulky items for a fee of £15 for up to three items. Bristol has one of the highest recycling rates of any city in the UK. In 2012, the residents of Bristol recycled 50% of their waste. This has steadily increased from a rate of 12% in 2004. This is due to the council providing home owners with kerbside recycling which takes all of their recyclable waste. This saves the residents having to go to the waste disposal plants and therefore encourages people to recycle more.
Transport	**Car sharing** Bristol council has set up a car club where you can hire a car from nearby whenever you need one. The council also has a car share page on their website. Some employers have a system for car sharing within their company; in many cases, car sharers are given priority parking, known as car club parking spaces. Bristol has 2+ people lanes for cars. This means that, at certain times during the morning and evening rush hour, only cars with two passengers can use these lanes on the road. **Walking** Bristol council has a partnership with walkit.com, a website which provides easy-to-read maps between any two points in Bristol. It displays the journey distance, the walking time at different paces, the number of calories used and the amount of CO_2 saved. **Public transport** Most of the major roads around the city now have a bus lane which cannot be used by private vehicles. Regular buses go into the city centre and the council is working towards producing a system very like the Oyster card system in London. **Cycling** Bristol has many cycle routes and became the UK's first cycling city in 2008. The government gave Bristol council £11.4 million to create dedicated cycle lanes, better facilities for bike users and more training for children. It created a dedicated cycleway which links the suburbs with the city centre. It has also provided many facilities for people who choose to cycle to work, with 300 cycle parking spaces in the city centre. The council has recently invested another £35 million in a plan to get a fifth of all commuters on their bikes by 2020, with the segregated cycle path numbers rising from 9% to 20% of the road area. There will be commuter corridors heading north, east, northwest and south from the city centre.
Affordable and energy-efficient housing	Houses are responsible for 25% of the UK's carbon footprint. It is therefore important that the council does what it can to improve energy efficiency in housing in order for Bristol to be more sustainable and to improve the quality of life for its residents. All new developments need to submit a sustainable energy strategy to the planning committee before they can get planning permission. Grants are available for loft insulation.

Figure 6 **Strategies to make urban living more sustainable and improve quality of life for the residents of Bristol**

Now test yourself

TESTED ☐

Explain how Bristol council has tried to improve the quality of life for the people who live in Bristol.

Exam tip

Select the information on quality of life that you want to remember and produce some sticky notes with the information.

9 Case study of a major city: São Paulo, Brazil

The context of the city of São Paulo influences its functions and structure

> **Site** is the land that the settlement is built upon.
>
> **Situation** is where the settlement is compared to physical and human features around it.
>
> **Connectivity** is the way that São Paulo is connected or linked to other settlements in Brazil and to other countries in the world.

> **Residential** is when an area is used for housing.
>
> **Public buildings** are buildings owned by the council that serve the residents of the city, such as a library.

What is the national, regional and global context of São Paulo?

REVISED

- National context – São Paulo is in the southeast of Brazil.
- Regional context – São Paulo is in Brazil, which is in the centre, east of South America with Paraguay to the west.
- Global context – São Paulo lies east of the Pacific Ocean and west of Africa; North America is to the north of Brazil.

Site, situation and location of São Paulo

REVISED

São Paulo is sited on a hilly plateau approximately 820 metres above sea level over which flow a number of rivers. The main city area is divided in two by the Anhangabaú River which now flows underground.

Brasilia
↑
1000 km

Rio de Janeiro
350 km ↗

Curitiba
← 330 km

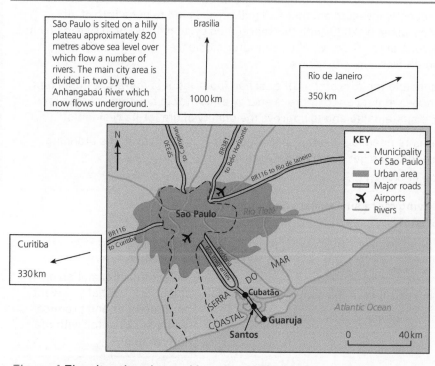

KEY
- - - Municipality of São Paulo
▓ Urban area
▭ Major roads
✈ Airports
Rivers

0   40 km

Paulistanos is the name given to the residents of São Paulo.

São Paulo metropolitan area is the whole of the built-up area; it includes São Paulo and a number of nearby cities and has approximately 19 million inhabitants.

São Paulo city area is the inner built-up area of São Paulo which has approximately 11 million inhabitants.

Figure 1 The site, situation and location of São Paulo

Now test yourself and exam practice answers at **www.hoddereducation.co.uk/myrevisionnotes**

What are the functions and building age of different parts of São Paulo's structure?

Cortiço is inner-city accommodation for the poor. Families live in one room with shared toilets and cooking facilities. Many of the buildings were previously office blocks or the homes of the wealthy before they left the inner-city area.

Favela is an area of homes for the poor, which can be found anywhere in the city. They are made from waste materials with no water supply, electricity or toilets.

Boulevards are wide, tree-lined streets.

Area of city	Functions	Building age
Central Business District (CBD)	• Businesses • Financial institutions • Shops • Hotels • Cultural establishments including museums, theatres and restaurants • Many high-rise residential blocks for wealthy residents • Main railway stations.	• Many buildings were constructed in the nineteenth century. • Rapid industrialisation in the twentieth century meant many high-rise multi-storeyed buildings were built both as office blocks and as homes for the wealthy.
Inner city	• Residential: wealthy regions, such as the Jardins; favelas, such as Paraisópolis, where 43,000 people are crammed into an area of 150 hectares.	• Nineteenth and early twentieth centuries.
Suburbs	• Residential: Morumbi, an area of high-security housing complexes, with a number of parks and shopping centres; favelas, such as Paraisópolis • Commercial • State government headquarters • University • Headquarters of banks • Multinational companies • Congonhas Airport.	• Twentieth-century developments.
Urban-rural fringe	• Residential – favelas and gated communities.	• First development of favelas in the 1980s. • Since 2000, development of gated communities.

Figure 2 **Functions and building age of different parts of São Paulo's structure**

Exam practice

Compare the functions of the CBD with that of the rural-urban fringe. (3 marks)

ONLINE

Exam tip

Remember, don't try to learn all the functions! Just learn a couple for each of the city areas.

São Paulo's character is influenced by its fast rate of growth

International migration is the movement of people from one country to another with the intention of staying there for at least a year.

Rural to urban migration is the movement of people from the countryside to towns and cities.

Push factors are the reasons why people want to leave rural areas.

Pull factors are the reasons why people are attracted to São Paulo.

National migration is the movement of people from one area of a country to another with the intention of staying there for at least a year.

Natural increase is when there is a positive difference between the birth rate and the death rate.

What are the reasons for the past and present trends in population growth?

REVISED

International migration

São Paulo has experienced many flows of international migration during the nineteenth and twentieth centuries. The first settlers were Portuguese but the main ethnic group in the city now is Italians. There are people from other European countries as well as from Asia, Africa and other parts of North and South America living in São Paulo, all drawn to the city by the high rates of economic growth that it experienced in the 1950s and 1960s. Now one-fifth of the population of the city originates from other countries.

Reasons for national migration

Push factors	Pull factors
In Brazil, 31% of rural households have no land. They have to rent land or find work as labourers, and as farms become more mechanised, there is the risk of losing their jobs. There is very little to keep people in the rural areas, so they move to the cities in search of work.	Infant mortality is higher in the rural areas (175 per 1,000) than in the favelas of São Paulo, where it is 82 per 1,000.
In the 1950s and 1960s there was a shortage of labour in São Paulo due to rapid economic growth of 226%. Advertising campaigns were run in the rural areas to attract workers to the city.	Word sent back to the villages by successful migrants makes life in the cities seem much better than it actually is.
Bahia in northern Brazil is very poor and periodically experiences drought. It has been estimated that 32 million people have chronic malnutrition in this region.	The rural dwellers have high expectations of a better quality of life in the city. There are more schools and doctors as the government puts more money into services for urban areas.
Land in rural areas has been taken from the subsistence farmers who were renting it from large landowners. These landowners now want to use the land to grow cash crops such as coffee and orange juice. Just eighteen landowners control an area six times the size of Belgium.	Rural to urban migration has slowed down in Brazil, although there is still migration between urban areas.

Figure 3 **Reasons for past and present population growth of São Paulo**

Exam practice

Explain **one** push factor and **one** pull factor of national migration.

(4 marks)

ONLINE

High rates of natural increase

The growth in population in São Paulo over the past 20 years has been caused by a high natural increase in population. The death rate has declined with improvements in health care, diet and housing conditions but the birth rate is still high in some parts of the city. The growth rate has, however, started to slow down from 5 per cent in 1975 to 1.3 per cent in 2013.

What is the impact of national and international migration on different parts of São Paulo?

REVISED

- São Paulo is culturally diverse due to the great variety of ethnic groups; for example, Bela Vista is known for its Italian culture and Liberdade for its Japanese culture.
- The large number of migrants has resulted in a young age structure and a high birth rate.
- Favelas have developed as the city has failed to keep up with the demand for housing.
- Other services, such as schools and hospitals in poorer neighbourhoods, are also unable to keep up with the growing demands.

Now test yourself

TESTED

Explain the impacts of migration on São Paulo.

Why is the growth of São Paulo accompanied by increasing differences in the quality of life of its residents?

REVISED

The process of deindustrialisation has been happening in São Paulo since the 1980s, causing a bigger gap between the rich and the poor in the city. This has caused an increase in the unemployment rate, which is at its highest for many years.

The growth of the population has been so rapid that the city has been unable to build sufficient housing for all the people. Therefore, the people who have jobs or have been in the city for a longer period of time tend to be the ones that are better off.

Managing the challenges of rapid growth in São Paulo

What are the effects of rapid urbanisation?

Favelas
The rapid industrialisation in the1980s caused the growth of favelas across the city due to the acute housing shortages. The areas did not have proper sewerage systems. Much of the sewerage runs down the streets into the rivers, people access water from stand pipes which serve hundreds of people.

Effects of rapid urbanisation

Traffic congestion and pollution
The residents of São Paulo own 6.2 million cars and there are 16,000 buses on the road. At times, there can be hundreds of kilometres of grid-locked roads. There is also the problem of the pollution that so many vehicles cause.

Unemployment
São Paulo could not provide jobs for all of its migrants which led to high unemployment rates of 19% in 1998 but reduced to 11% in 2012.

Now test yourself

What are the effects of rapid urbanisation on São Paulo?

TESTED

What are the advantages and disadvantages of both bottom-up and top-down approaches to solving São Paulo's problems?

Scheme	Advantages	Disadvantages
Bottom-up approach – self-help scheme in Santo Andre The aim of the scheme was to improve infrastructure and services. Improvements include: ● Health care has been made more available ● Literacy courses made available for adults ● Recreational facilities made available ● Favelas have been upgraded ● Credit facilities have been made available to small-scale entrepreneurs.	● Community is included in the decisions that are made. ● Housing in the area will be the same type of housing but will be more substantial and will have services. ● The improvements are not just housing but help to further improve the quality of life of people in the area.	● Schemes take a long time to instigate. ● With so many different people involved, it is hard to get agreement on how the money available should be spent. ● It is difficult to get people to accept help with the literacy programmes.
Top-down approach – Cingapura Housing Project 14,000 new homes were built. The project began in 1995 and ran until 2001. The new homes were built in blocks about 10 storeys high. The favela residents were then expected to pay a rent of about US$26 a month for their new apartment.	● The housing had clean water supply and proper sanitation. ● The new housing was built on the same land as the favelas, so people did not have to leave the area they knew. ● Leisure areas were included in the developments.	● Many favela owners have never paid rent and can't afford to. ● Favelas were demolished to build the new blocks. ● There was no provision for small businesses. ● The type of accommodation is forced on the inhabitants who have no say in what is being built. ● The living space in each apartment was very small.

Figure 4 **Advantages and disadvantages of bottom-up and top-down approaches to solving São Paulo's problems**

What is the role of government policies in improving the quality of life in São Paolo?

REVISED

The government has attempted to improve the quality of life for the people who live in São Paulo in a number of ways:

- A government bank (BNH) has funded housing projects and provided low-interest loans to lower- and middle-income people to help them to buy a home.
- A scheme which built houses for teachers and other people who worked for the government.
- A scheme to build government-owned housing, which also funded self-help projects to upgrade housing in the favelas.
- An underground train system, which was built in the 1960s and opened in 1974. There are now six lines and 65 stations carrying three million passengers a day.
- The city has also instigated busways (similar to bus lanes in the UK) in an attempt to deal with traffic problems. The buses have sole use of these lanes.

Bottom-up approaches are self-help schemes; the residents of an area are in charge of what happens. The government usually provides monetary help and advice on how to improve their housing.

Top-down approaches are when the government improves an area and expects people to move in to the housing they have provided. Sometimes the government borrows large sums of money to pay for the scheme.

Now test yourself

TESTED

Explain how government policies have tried to improve the quality of life in São Paulo.

Exam practice

Define the term 'top-down approach'. (1 mark)

ONLINE

10 Global development

Definitions of development vary as do attempts to measure it

Secondary sector includes manufacturing industry. The number of people employed in this sector increases as a country develops.

Primary sector is made up of extractive industries, such as farming, fishing, forestry and mining. Developing countries have high numbers of people employed in this sector.

Development refers to an improvement in the quality of life for the population of a country.

Tertiary sector is the service industry and includes jobs such as teaching. Very few people are employed in this sector of industry in a developing country.

Quaternary sector includes financial services and telecommunications.

Development gap is the difference between the parts of the world that have the wealth and the parts that do not.

North–South divide is a virtual socioeconomic and political line on the globe which splits the developed and wealthy countries in the 'North' from the poorer developing countries the 'South'.

What are the contrasting ways of defining development, using economic criteria and broader social and political measures?

REVISED

Economic development	An increase in a country's wealth. This could be an increase in people working in the secondary sector and a decrease in the numbers of people working in the primary sector. It could be indicated by a greater use of natural resources, for instance, energy use per head of population increases.
Social development	A number of changes that have a direct impact on the population's quality of life. This could include improved levels of literacy through greater access to education, better housing conditions and more doctors.
Political development	Freedom for the people to have a greater say in who governs their country.
Cultural development	This could involve better equality for women and better race relations.

Figure 1 **Four types of development**

Exam tip

Don't forget to learn the key words for this section.

Now test yourself

Describe the different types of development.

TESTED

What factors contribute to the human development of a country?

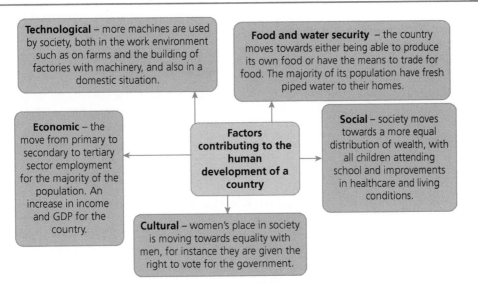

Technological – more machines are used by society, both in the work environment such as on farms and the building of factories with machinery, and also in a domestic situation.

Food and water security – the country moves towards either being able to produce its own food or have the means to trade for food. The majority of its population have fresh piped water to their homes.

Economic – the move from primary to secondary to tertiary sector employment for the majority of the population. An increase in income and GDP for the country.

Factors contributing to the human development of a country

Social – society moves towards a more equal distribution of wealth, with all children attending school and improvements in healthcare and living conditions.

Cultural – women's place in society is moving towards equality with men, for instance they are given the right to vote for the government.

Exam practice

State **two** factors that contribute to the human development of a country. (2 marks)

ONLINE

What are the different ways that development is measured?

REVISED

Development measure	Description
Gross domestic product (GDP) per capita	This is the value of all the goods and services produced in a country during a year, in US dollars. Per capita means that the figure is divided by the number of people who live in the country to give an average per person.
Human Development Index (HDI)	This is a comparative measure of different aspects of life between countries. The measures used are life expectancy, education and standards of living.
Measures of inequality	These are ways of measuring how equal people are within a country or between countries. This is often a measurement of the wealth or health care of people in a country or between countries.
Corruption Perceptions Index	This is the perceived corruption in governments and the public sector. It is a perception because corruption is hidden and therefore difficult to measure. It means that government officials are using development for their own betterment rather than the betterment of the country.

Figure 2 **The different ways that development is measured**

Exam practice

Define the term 'Human Development Index' (HDI). (2 marks)

ONLINE

The level of development varies globally

Gross national income (product) (GNI) is the value of all the goods and services produced in a country and from its exports during a year, in US dollars. The per capita means that the figure is divided by the number of people who live in the country to give an average per person.

Tectonic activity is the movement of the Earth's plates.

Mineral is a solid, naturally occurring inorganic substance.

Fossil fuel is a fuel such as oil or gas, which is naturally occurring and formed from the remains of dead organisms.

What is the global pattern of development and how is it uneven between and within countries, including the UK? REVISED

The level of development in a country depends upon the measure that is being used to assess the development. For example, if GNI is the measure being used, Sweden and France are in the top category but they are not if HDI is the measure being used.

When individual countries are considered, most have areas which are richer than other areas. This is also true in towns and cities. For example, the Bronx in New York has very poor areas but many people consider New York to be a wealthy city with a high HDI. Therefore, using broad country figures hides a lot of unevenness within countries.

Using the UK as an example, the country is split between the north and the south, the income in the south and east being far higher than that in Yorkshire and Lancashire. This still hides the true picture however, because not all of the people who live in the south and east earn a high income.

What are the factors that have led to spatial variations in the global level of development? REVISED

Factor	Description
Physical	Climate – countries which have average rainfall and moderate temperatures are able to support their populations with the food that they produce.
	Landlocked countries – countries which do not have a coastline find it difficult to trade their goods because they have to rely on the goodwill of their neighbours to allow them to drive their products to the coast and for them to receive imported goods.
	Natural hazards – floods, tectonic activity, droughts and hurricanes are more likely to occur in some countries than others. Many of the countries at risk of experiencing these hazards are developing because income on a regular basis has to be diverted to help recover from these events.
	Natural resources – resources such as minerals and fossil fuels help a country to develop because the extraction and sale of these resources will bring income into the country.
Historic	Colonies – supplied food to the country which 'owned' them. For example, Brazil sent food and minerals to Portugal. This hindered the development of the colony but aided the development of the owner country.
	Trade – many trading partnerships go back to colonial times. However, countries with good trading partners or countries on trade routes developed more quickly than countries which did not trade with other countries.
	Politics – countries with stable governments developed more quickly. If countries are at war or are going through civil wars, their income is spent on military weapons rather than on development. If a country is run by a dictator who is corrupt, development can be halted for most of the country as money is being spent on an affluent lifestyle for an elite group of people who rule the country.

Figure 3 **Factors that have led to spatial variations in the global level of development**

Factor	Description
Economic	Foreign direct investment (FDI) – this can help a country to develop because it brings money into a country. Countries in Africa receive 5% of foreign direct investment with 15% of the world's population. Countries in Europe receive 45% of foreign direct investment with 7% of the world's population. However, things are changing as companies from developed countries start to invest in emerging countries. For example, Coca-Cola in India. World trade – the developing countries sell primary products to developed countries. Manufactured goods are worth more money than primary products so developed countries earn more from their trade than developing countries. Infrastructure – this is the roads, railways and facilities, such as electricity. Developed countries have good infrastructure and therefore companies want to invest in them because they know their goods will be produced and moved quickly.

Figure 3 **Factors that have led to spatial variations in the global level of development**

Now test yourself

TESTED

Explain the factors that have led to variations in development globally.

What are the factors that have led to spatial variations in the level of development within the UK?

REVISED

Reason		Development outcome
Physical	Relief	The south of the UK is flatter; this aids development as urban areas can be easily built. The north and west are more mountainous, making urban areas and communications routes more difficult to build.
	Climate	The south and east of the UK have a better climate than the rest of the country with less rainfall. This makes it a more pleasant area to live in.
	Natural resources	The Midlands, the North and South Wales started to develop with the discovery of natural resources; in the first instance, this was the mining of coal.
	Position	The south and east of the country are closer to the communication links to Europe. This makes companies want to locate in this area.
Historic	Politics	The seat of the government is in London in the Southeast. This made it a highly desirable location for businesses in the past as they were close to where decisions were being made and found out about them quickly.
	Colonies	Although ships sailed for the colonies from ports on the west of the country, all the decisions were taken in London on the east of the country.
Economic	Infrastructure	The infrastructure in the London area is the best in the country. All roads lead to the centre of London. Companies who located there would be able to trade with the rest of the country easily.
	Foreign direct investment (FDI)	Most foreign direct investment into the UK is in London although the government has tried to encourage investment in other areas, for example Honda in Swindon.

Figure 4 **Reasons for the UK's uneven development**

Revision activity

Study the reasons for the uneven development of the UK. Draw a spider diagram to help you remember the information. Don't forget to add some examples.

Exam tip

Giving examples of countries would help in a question on spatial variations of development.

Consequences of urban global development

What is the impact of uneven development on the quality of life in different parts of the world?

REVISED

Access to housing
In developing countries, a large percentage of the population live in poor quality houses without proper water or sanitation. In developed countries people live in houses with a fresh water supply and sanitation.

Health
The lower the development levels in a country, the higher the number of people per doctor. As a country develops, the number of doctors increases with the increase in levels of education.

Food and water security
As a country develops, it becomes more secure in its food and water supply, partly because it has the technology to improve the intensity of agricultural production and to provide water supplies, but also because it has the wealth to buy in food if it cannot produce enough itself.

Impact of uneven development on quality of life in different parts of the world

Education
Literacy levels relate directly to the level of development of a country. The lower the country's GDP, the lower the country's literacy rate.

Technology
As a country becomes more developed, the level of technology in the country improves. One way of measuring this is by looking at the internet users in the country.

Employment
In developed countries, the majority of the population work in tertiary and quaternary industry. In developing countries, the majority of the population work in primary industry.

Literacy is the ability to read and write.

Employment structure is the numbers of people employed in each sector of industry.

Anomaly is something that is outside the norm.

World Bank is an international financial institution that provides loans to developing countries.

Donor countries are countries that give aid.

Exam practice

Describe how uneven development has an impact on the quality of life in different parts of the world.

(3 marks)

ONLINE

Exam tip

This question would be enhanced with the use of examples.

Strategies to address uneven development

What international strategies attempt to reduce uneven development?

REVISED

Bilateral aid	The first aid that was made available to countries was after the Second World War. This was aid from the USA to Europe to help rebuild its cities after the intense bombing that had taken place. This aid is known as bilateral aid and is given from one government to another government, usually with attached agreements, such as the recipient country has to give building projects to the donor country. An example of this is the money that India has loaned to Bhutan to build HEP schemes. India has provided the engineers and the technology and will get the electricity that is produced at a cheaper rate than the local people.
Multilateral aid	This is when developed countries give money to international organisations such as the World Bank or the United Nations. These organisations then redistribute the money in the form of loans to poorer countries.
Official and voluntary aid	Governments such as the UK and USA provide money which charity organisations can bid for to develop aid projects in different countries of the world.
Voluntary aid	This is the money raised from donations and charities. Organisations such as Oxfam and Save the Children raise money through fund-raising events, private donations and charity shops. This money usually funds bottom-up projects.
Inter-governmental agreements	These are agreements between developed world nations that work together to provide aid for developing countries. One of these agreements is with the EU. The EU delivers aid in different ways; one of these is sector support. For example, the EU provides funds to develop education in a country. This means that the EU would provide funds directly to the education department of the partner country. The development of emerging countries such as Brazil, China and India has seen these countries become major donor countries of aid. This is known as the South-South development co-operation, which has 30 countries that are donor members who help to develop countries that are not so well off as they are. One of the biggest donors is China although much of the aid given by China does have strings attached.

Figure 5 **Strategies that attempt to reduce uneven development**

Now test yourself

TESTED

Make a list of the international strategies that have been used to reduce uneven development.

What are the advantages and limitations of top-down and bottom-up development projects?

REVISED

Top-down projects

Advantages	Disadvantages (limitations)
The country will develop more quickly because of the size of the project.	The country will go into debt. In some cases these debts have never been paid off. The end product is usually expensive to maintain.
The scheme is run by the government so is likely to achieve its development objectives.	The debt may have conditions attached to it which mean that the country is under external influences for many years.
In some cases, for example large HEP schemes, it is the only way to raise the capital due to the size of the project.	Much of the building work is done by machines or by companies from other countries, so local jobs are not created.
It is a way of helping the large urban populations of a country, but often at the expense of the rural areas.	Local people have no say in what happens. In many cases they have lost land.

Figure 6 **The advantages and limitations of top-down projects**

Bottom-up projects

Advantages	Disadvantages (limitations)
The scheme is run by the local people so is likely to achieve its development objectives as they decide what happens.	The country will develop more slowly because of the size of the project.
The scheme uses appropriate technology and the end product is usually cheap to maintain.	It does not help the majority of the population who live in urban areas.

Figure 7 **The advantages and limitations of bottom-up projects**

Appropriate technology is technology that is suitable for the skill level of the country it is in.

Top-down projects are development programmes that are initiated and run by the government on a large scale. The government borrows money from organisations such as the World Bank to finance a large-scale scheme to benefit the whole country, for example, the Three Gorges Dam in C hina.

Bottom-up projects are development programmes that are run by local community groups. This is development on a small scale. They are schemes that are planned and controlled by local communities to help their local area. They are not expensive because they use smaller appropriate technology. Local people fund the schemes themselves or with help from aid groups.

Exam practice

Explain **one** advantage and **one** disadvantage of bottom-up development projects. (4 marks)

ONLINE

The level of development of Tanzania is influenced by its location and context in the world

What is the location and position of Tanzania in its region and globally?

Tanzania is in eastern Africa. It has borders with Kenya and Uganda in the north, Rwanda, Burundi and the Democratic Republic of the Congo to the west and Zambia, Malawi and Mozambique to the south.

What is the broad political, social, cultural and environmental context of Tanzania in its region and globally?

Exam tip

Always know where your case studies are in the world. Create a world map with them on it to help you to remember their location.

- Tanzania is the thirteenth largest country in Africa.
- It has 800 kilometres of coastline.
- It contains Africa's highest mountain, Kilimanjaro. It is mountainous and densely forested in the northeast. Central Tanzania is a large plateau with plains and arable lands. The eastern shore is hot and humid. Three of Africa's great lakes are partly in Tanzania: Lake Victoria, Lake Nyasa and Lake Tanganyika. 38 per cent of its land area is set aside in conservation areas.
- The country has two main languages, English and Swahili.
- The former capital Dar es Salaam has most of the government offices. It is the country's largest city and port and is the country's wealthiest area.
- Dodoma became Tanzania's new capital in 1996. Dodoma is in the middle of the country and this was an attempt to try to improve the standard of living in that area.
- On the United Nations' Human Development Index in 2013 Tanzania was 152 out of 187 countries.
- The country was a German colony from 1885 until the end of the First World War, when it came under British rule. In 1967, the government of the country realised that wealth was not remaining in the country but was leaking to other countries that had investments in Tanzania. It adopted a socialist political and economic approach, and nationalised all of the banks and large industries.
- The first multi-party elections were held in 1995 when Benjamin Mkapa was elected president. The country returned to a free market economy and foreign direct investment was encouraged. The country is now trying to develop through building infrastructure and to reduce poverty as it is one of the poorest countries in the world.

Human Development Index (HDI) is a composite index measuring average achievements in three basic dimensions of human development – life expectancy, education and standards of living.

Socialism is a political model based on the belief that a country's people should own the means of production and regulate its political power.

Nationalised is when a business or industry is converted from private to government ownership

Periphery contains less wealth and, on some occasions, most of the population.

Core contains the most wealth.

How is development uneven within Tanzania and what are the reasons why development does not take place at the same rate across all regions?

- The fastest rates of development have taken place around Dar es Salaam, which used to be the capital of the country and still is the main port. People can get employment in industries related to the port.
- The agriculture sector has been slow to develop with many farmers still using traditional hand hoes and the agriculture is rain fed with no irrigation. Yields are low and 80 per cent of the arable land is used by smallholders who do not have the means to improve their farms.
- Many of the regions have used the internet to try to attract development. The local government has produced professional booklets about the areas to attract foreign direct investment.
- There has been development in some areas with export crops of sugar, tea and tobacco doing well.
- Some of the areas are very remote and this is another reason why development has been slow. They are on the periphery of the country and do not have the infrastructure to take crops to market for sale.
- Other parts of the country, which are in the core area along the Indian Ocean coastline, have developed more quickly as they have better opportunities. Areas with gold deposits have also seen an improvement in their incomes as investors from other countries have started to extract the mineral.

The interactions of economic, social and demographic processes influence the development of Tanzania

Transnational corporations (TNCs) are large companies that have their headquarters in one country and branches all over the world.

Public investment is money put into businesses by the government.

Private investment is money put into businesses by TNCs and smaller businesses.

Informal sector refers to people who set up businesses such as selling products on the street; they do not pay taxes or rent proper business premises.

Balance of trade is the difference between a country's imports and exports.

International Monetary Fund (IMF) is an international financial institution that was set up in 1945 to promote international trade. A fund is maintained from member nations who can then make withdrawals when their economy is in difficulty.

What are the positive and negative impacts of changes in the sectors of Tanzania's economy?

REVISED

Sector	Positive impacts	Negative impacts
Primary	• Aid has been given to farming communities to try to introduce irrigation techniques which use appropriate technology to raise crop yields. • Mining of natural resources has brought much needed foreign direct investment into the country. • The recent discovery of gas and oil will also help to provide much needed foreign direct investment in the future.	• Agriculture's share of GDP fell from 29% in 2001 to 24% in 2010. • Tanzania uses an average of 9 kg of fertiliser per hectare, whereas Malawi, at a similar stage of development, uses 27 kg and China 279 kg. • The sector is also still dependent on the weather and dry years mean low yields from crops.
Secondary	• This sector's share of GDP increased from 18% in 2001 to 22% in 2012. This will provide extra money for the economy.	• Manufacturing's share of GDP is fairly constant with a slow growth rate. It is concentrated on a few goods which are of low value. Only 5% of employed people work in this area. • This sector needs to develop other goods through foreign direct investment to provide employment for people who are low on skills.

Figure 8 **Tanzania's economy by sector**

Sector	Positive impacts	Negative impacts
Tertiary	● The service sector continues to grow with the development of a small middle class. Its growth rate was 8% between 2001 and 2012. ● With more people in higher paid jobs, the country should receive more taxes to help it develop. ● There has been a growth in employment in education with the expansion of primary school education to all children. ● There has also been an increase in health care workers, providing employment for people who acquire the skills.	● If the country is to develop, the service sector needs to continue to grow. ● Many of the jobs in the tertiary sector require a level of skill which will only be acquired through education.
Quaternary	● The communications and financial services sector are the fastest growing in the economy, growing 15% between 2003 and 2012.	● The sector requires a highly skilled workforce, which receives high wages. Therefore, it does not provide employment for the low-skilled Tanzanian workforce, so its impact on the reduction of poverty is low.

Figure 8 **Tanzania's economy by sector**

What are the characteristics of international trade and aid and how is Tanzania involved in both?

REVISED

International trade	International aid
Tanzania's largest trading partners in 2012 were South Africa, Switzerland and China, with exports worth US$5.5 billion. However, its imports totalled $11.7 billion and its main import trading partners were Switzerland, China and United Arab Emirates. This meant the country had a negative balance of trade. The country has strong telecommunications and banking sectors, and its tourism sector continues to grow. It contributed 12.7% of Tanzania's GDP and employed 11% of the workforce. The economy, however, is still based on agriculture, which makes up 24.5% of GDP and 85% of exports. Industry and construction are growing, providing 22.2% of GDP in 2013. This includes mining, manufacturing, electricity, gas and water supply. The country has started to extract diamonds, tanzanite and gold to export. The coal which is extracted is used domestically.	Tanzania is the second-largest aid recipient in Sub-Saharan Africa, after Ethiopia. They received a combined total of US$26.85 billion in aid between 1990 and 2010. The main donors of aid to Tanzania are the USA and the UK who gave US$793 and US$307 million respectively in 2013. Other countries that give substantial amounts are Japan, Canada, Germany and Norway. Aid has been given to support the general budget of the country, but also to fund education, health care provision, water supply and sanitation.

Figure 9 **Tanzania's involvement in international trade and aid**

What is the changing balance between public investment and private investment in Tanzania?

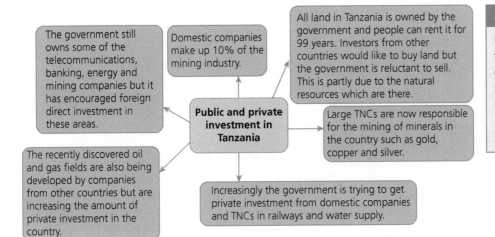

The government still owns some of the telecommunications, banking, energy and mining companies but it has encouraged foreign direct investment in these areas.

Domestic companies make up 10% of the mining industry.

All land in Tanzania is owned by the government and people can rent it for 99 years. Investors from other countries would like to buy land but the government is reluctant to sell. This is partly due to the natural resources which are there.

Public and private investment in Tanzania

Large TNCs are now responsible for the mining of minerals in the country such as gold, copper and silver.

The recently discovered oil and gas fields are also being developed by companies from other countries but are increasing the amount of private investment in the country.

Increasingly the government is trying to get private investment from domestic companies and TNCs in railways and water supply.

Revision activity

Use the spider diagram to make two lists: government investment in Tanzania and private investment in Tanzania. Remember, you don't have to learn everything!

What are the changes in population structure and life expectancy that have occurred in the last 30 years in Tanzania?

- Total population was just over 51 million in 2015.
- Approximately 70 per cent of the population live in rural areas.
- Approximately half of the population is under the age of 15.
- Life expectancy is improving but is still relatively low.

Exam practice

Suggest **two** reasons why the population structure of Tanzania has changed over the last 30 years.

(4 marks)

What are the changing social factors in Tanzania?

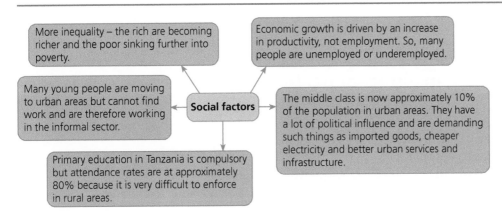

More inequality – the rich are becoming richer and the poor sinking further into poverty.

Economic growth is driven by an increase in productivity, not employment. So, many people are unemployed or underemployed.

Many young people are moving to urban areas but cannot find work and are therefore working in the informal sector.

Social factors

The middle class is now approximately 10% of the population in urban areas. They have a lot of political influence and are demanding such things as imported goods, cheaper electricity and better urban services and infrastructure.

Primary education in Tanzania is compulsory but attendance rates are at approximately 80% because it is very difficult to enforce in rural areas.

Changing geopolitics and technology impact on Tanzania

How do geopolitical relationships with other countries affect Tanzania's development?

Geopolitical relationship	Effect
Foreign policy and military pacts	Tanzania has never had a civil war but has been involved in other countries' disputes through its foreign policy which, on some occasions, has been costly. • The anti-Amin forces in Uganda in 1978 cost the country approximately US$500 million. • In the past, Tanzania has hosted refugees from neighbouring countries including Mozambique, Burundi and Rwanda, usually in partnership with the United Nations.
Defence	The country has a small army, navy and air force. Their main work is being part of United Nations' peace-keeping missions to other countries, such as the Lebanon and Sudan. There are 25,000 regular personnel and 80,000 reserves.
Territorial disputes	Tanzania is involved in an ongoing dispute with Malawi over the ownership of Lake Nyasa. The lake covers 30,000 sq km and, according to Malawi, was given to Malawi in the 1980 Heligoland-Zanzibar Treaty between Germany and the UK. Tanzania argues that the boundary between the two countries is in the middle of the lake. This means that Tanzania owns half of the lake and Malawi the other half. The dispute has been going on for years but was restarted in 2012 when Malawi gave a British company the right to oil exploration in the lake; to date the dispute has not been solved.

Figure 10 **The effects of geopolitical relationships on Tanzania's development**

> **Geopolitical** refers to the influence of factors such as geography and economics on the politics and foreign policy of a state.
>
> **Refugee** is a person who has been forced to leave their country in order to escape war, persecution or a natural disaster.

How does technology and connectivity support development in different parts of Tanzania and for different groups of people?

- The government of Tanzania has invested money into producing an ICT network for the whole country.
- This will provide the necessary fibre cables for other network providers, such as mobile phone companies and broadband suppliers, to supply people in their homes.
- Mobile usage in Tanzania has increased greatly over the last decade with nearly 60 per cent of the population having mobile phones and many of them using the internet via their mobiles.

Positive and negative impacts of rapid development in Tanzania

What are the positive and negative social, economic and environmental impacts of rapid development for Tanzania and its people?

REVISED

	Positive	Negative
Social	• Improvement in life expectancy. • Improvement in supplies of fresh drinking water and sanitation; 62% of the population now have supplied water and sanitation. • All children have access to primary schools and attendance is above 80%.	• Some rural areas are not benefitting from improvements at all. • In 2012, 28% of the population still lived below the poverty line. • Because of the rapid expansion in schools, teaching standards are low; 60% of students failed the secondary school leavers' exam in 2012. • Health care is still poor with approximately 40% of the jobs not filled because of a lack of health care professionals in the country.
Economic	• Reduction in poverty. • Improvements in GDP for the country. • Foreign direct investment in the country is improving • Strong banking, financial and telecommunication sectors.	• There is still inequality between regions. • There are indications that there is still a large divide between rich and poor. In 2012, the most affluent 20% of the population accounted for 42% of total consumption whereas the least affluent 20% consumed only 7%.
Environmental	• With the introduction of electricity to rural areas using bottom-up schemes, deforestation will slow down. • Proper irrigation schemes using appropriate technology will allow the farmers to use their land more efficiently and stop overgrazing.	• The extraction of minerals in some areas has caused environmental problems because quarries are left as scars on the landscape. • Deforestation caused because of rises in population numbers and the use of wood fuel for domestic purposes. • Deforestation leads to loss of habitats and biodiversity. • Overgrazing of farms in dry years is also a problem. • Gold mining causes problems of toxins leaking into water courses.

Figure 11 **Impacts of development on Tanzania**

Revision activity

Draw a table which shows the positive and negative social, economic and environmental impacts of rapid development for Tanzania and its people. Select the information from Figure 11 that you want to remember.

Deforestation is the cutting down and removal of all or most of the trees in a forested area.

Overgrazing is when grass is grazed so heavily that the vegetation is damaged and the ground becomes liable to erosion.

How are Tanzania's government and people managing the impacts of rapid development to improve quality of life and its global status?

- Aid from donor countries has provided primary education for all.
- The country's infrastructure has improved.
- The majority of the population, even in urban areas, still do not have access to piped drinking water with the figure standing at approximately 34 per cent; the figure for proper sanitation being approximately 12 per cent. Some of these problems have been caused by the rapid population growth, but if the country is to develop, the government must deal with these problems.
- The quality of life in Tanzania has not really improved for the majority of the population, who still live in rural areas and work in agriculture. The government is trying to help in these areas but development is slow and quality of life for the majority is not really improving.
- Tanzania's status in the global community is in some ways above that of its neighbours as it has never had a civil war and has indeed helped the UN on a number of occasions with refugees from countries that have internal problems.

Now test yourself

State **three** ways that Tanzania's government and people are managing the impacts of rapid development.

Exam tip

Tanzania is a big case study. You will need to divide the material into manageable chunks under the headings given. The exam questions will focus on the headings that are used in the text.

11 Resource management

A natural resource is any feature or part of the environment that can be used to meet human needs

How can natural resources be defined and classified in different ways?

- **Biotic factors** are all the living things that are found in an area.
- **Abiotic factors** are all the non-living things in an area such water, wind and oxygen.

- **Renewable resources** can be used again and will not run out, such as the Sun or water.
- **Non-renewable resources** are finite in their supply. This means that there is a limit on the supply of the resource, such as coal.

> **Exam practice**
>
> Explain **two** ways in which people exploit the environment. (2 marks)
>
> ONLINE

What are the ways in which people exploit environments in order to obtain water, food and energy?

	Exploitation	Environmental changes
Water	Fresh water is a resource that is needed for people to survive. As the number of people in the world continues to increase, the need for water will also increase. Water is used for many things including drinking, washing and producing manufactured goods. In many countries, it is not these uses that exploit water but the misuse of water sources, for example the extraction of minerals.	In many countries, including the UK, ground water is being used faster than it can be replenished by rain. This causes problems for plants and animals, and could cause a decrease in biodiversity in areas. When minerals are extracted, toxic by-products can be washed into rivers causing a decrease in the quality of the water that is used for human consumption in the area.
Food	Farming – with increased population numbers in many countries, farming land is being overgrazed. Fishing – in many areas, overfishing has occurred due to the demand being so great that fish stocks cannot replenish themselves quickly enough.	If land is overgrazed, the bare soil is left exposed to the weather. The rain and wind can cause the soil to be eroded and either washed or blown away. Overfishing has led to a reduction of biodiversity in the oceans. As the ocean is a balanced ecosystem, if some fish species are reduced, it has an impact on the whole ecosystem.
Energy	The extraction of fossil fuels to produce energy can cause a number of problems. The reserves of fossil fuels, such as oil and natural gas, have been dramatically reduced because of this exploitation, although there are still large reserves of coal.	The burning of coal to produce energy in the UK has caused acid rain to fall in Norway and Sweden. This has caused trees in forests to die, resulting in a reduction in biodiversity.

Figure 1 Changes caused by the exploitation of resources

The patterns of the distribution and consumption of natural resources vary on a global and a national scale

What is the global variety and distribution of natural resources?

REVISED

Soil and agriculture

Soil regions of the world are in very broad sweeping bands across countries and continents. The type of soil also relates to the climate and vegetation of the area. Some soils, such as chernozems and brown forest soils, are fertile and correspond with areas of high agriculture production, whereas other soils are less fertile. Agriculture in these areas is less productive.

Forestry

Forestry on a global scale is concentrated in certain areas. Countries which produce at least five per cent of the world's wood production include Canada, Brazil and the USA. Most countries have their own forestry industry.

Fossil fuels

The countries that have the most oil reserves are Venezuela, Saudi Arabia and Canada. The countries with the highest gas reserves are Russia, Iran and Qatar. The USA has the highest reserves of coal left in the world although Russia and China also have vast reserves.

Rocks and minerals

Rocks can be igneous, sedimentary or metamorphic and are distributed around the globe. These main categories can be broken down into hundreds of different types of rocks. The most common rock type on the surface of the Earth is sedimentary. This layer of rock is very thin and goes about 2 km down into the crust. Below this level the most common rocks are igneous and metamorphic.

Minerals are distributed across the world, although some are concentrated on certain continents; for example, diamonds, which are precious minerals, are found in Sub-Saharan Africa, Russia and Australia but have not yet been discovered on other continents. Other minerals, such as iron ore, are distributed on all continents except Africa.

Water supply

The water that is used by humans is from rainfall, rivers, ground sources and, in some cases now, from the sea. However, the concern is that there will not be enough water for a global population expected to reach eight billion by 2025. In some countries, expensive water transfer schemes have already been built because the available water supply is not where the majority of the people live.

Distribution is how something is shared out or spread across an area.

Natural resources are materials that are provided by the Earth that people make into something that they can use.

Arable farming is the growing of cereal crops

Pastoral farming is the breeding of sheep, cattle, pigs or any other animal on a farm.

Fossil fuel is a fuel that formed from the remains of living organisms.

Mixed farming refers to a farm which has cereal crops and animals.

Exam tip

You should be aware of the global distribution of natural resources. Try to learn a major country for each of the resources, for example, Canada for forestry.

What is the UK variety and distribution of natural resources?

Soil and agriculture

The UK has varied soils, many of which are very fertile. This means that UK farmers have a wide choice of crops or animals that they can farm. Many different kinds of arable and pastoral farming are practised. Some farmers have also started to grow vines and British wine is now being produced in Kent, Sussex and Devon. In Cornwall, farmers have even begun to grow and make British tea.

Forestry

Woodlands are distributed across the UK. Some of the woods are in private ownership and others are owned by the Forestry Commission. The Forestry Commission was set up because of the vast amount of deforestation that had occurred in the UK. The country now has a healthy forestry industry which employs approximately 800,000 people. It makes up approximately 2.5 per cent of the British economy each year. Forestry is concentrated more in the north and the west of the country where the land and climate is less agreeable and therefore more difficult to farm.

Fossil fuels

Fuels such as coal, oil and gas are all found in the UK. Coal has been mined in the country for hundreds of years and was the first source of fuel to power steam engines. The coal reserves in the UK were vast but most were a long way below ground. Coal was found in South Wales, Kent and the Midlands (Staffordshire, Yorkshire, Derbyshire, Nottinghamshire). Coal reserves were also mined in Northumberland and Durham, Scotland and Northern Ireland.

Oil and gas have more recently been discovered under the North Sea, although the UK does have one inland oilfield at Wytch Farm on Poole Harbour.

Water supply

Although the UK does have a plentiful supply of rainfall, it doesn't tend to fall in the places which have the highest concentration of population (see Chapter 13). This means that certain areas of the UK can become short of water in the summer when demand is high, especially if there has been a dry spring.

Rocks and minerals

The UK has varied rock types which have already been discussed in Chapter 1. Other major rock types in the UK are clay and limestone.

The UK has a variety of minerals. They are used in construction to build houses and roads, as well as in industry, agriculture and horticulture. In 2013, 195 million tonnes of minerals were extracted from the UK landmass, which can be broken down into the following main categories:
- Construction minerals – 157 million tonnes
- Industrial minerals – 24.6 million tonnes
- Fossil fuels – 13.9 million tonnes.

Another 90.1 million tonnes of minerals were also extracted from under the sea (oil, gas, sand and gravel).

> **Exam tip**
>
> It would be a good idea to learn where some resources are found both globally and in the UK.

Exam practice

State **two** places in the UK where fossil fuels can be found. (2 marks)

ONLINE

What are the global patterns of usage and consumption of food, energy and water?

REVISED

Food consumption

The countries with the highest levels of food consumption are developed countries, with much of Europe being in this category. The countries with the lowest levels of food consumption per person are in developing countries, such as countries in Sub-Saharan Africa.

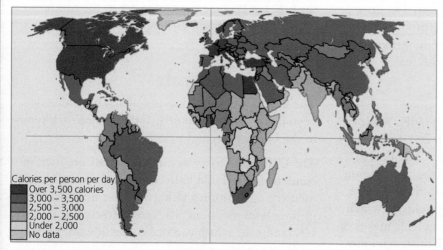

Figure 2 Daily food consumption, 2014

Now test yourself

TESTED

Use Figure 2 to describe the global calorie intake per day.

Energy usage

The amount of energy used by a country depends on many factors; one of these is the level of development of the country. Developed countries have a much higher demand for energy than developing countries. Emerging countries use large amounts of energy to power their developing industries. The demand for energy in the world continues to increase.

Exam tip

You may be asked to describe a distribution on a map.

- You should start with **general** points.
- Your answer should then become more **specific**.
- If data is asked for, you will lose a mark if you do not include it.
- If data is not requested, you should still include some as you will be given credit.

Exam practice

Study Figure 3. Suggest **two** reasons why the energy usage per continent varies greatly. (2 marks)

ONLINE

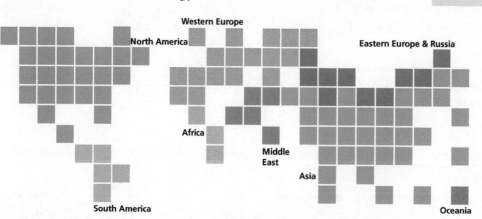

Figure 3 Energy consumption per region. Each symbol represents one per cent of world primary energy consumption

Water usage

The amount of water used varies greatly between different countries in the world. It also varies with the level of development of a country. The map in Figure 4 shows the amount of water that is available to be used in each country. The demand for water is dealt with in more detail in Chapter 13.

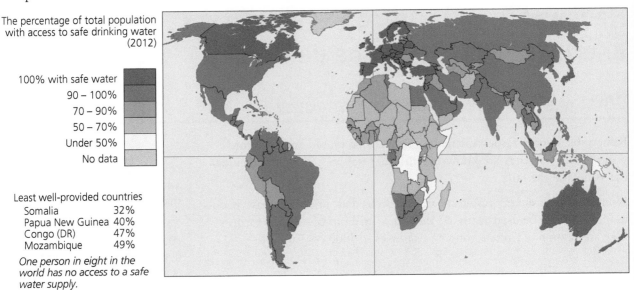

The percentage of total population with access to safe drinking water (2012)

100% with safe water
90 – 100%
70 – 90%
50 – 70%
Under 50%
No data

Least well-provided countries
Somalia   32%
Papua New Guinea 40%
Congo (DR)   47%
Mozambique   49%

One person in eight in the world has no access to a safe water supply.

Figure 4 Global water supply

> **Exam tip**
>
> Learn the global pattern of usage and consumption of food, water and energy. You may be asked a question relating to a map or a question which draws on your own knowledge.

12 Energy resource management

Renewable and non-renewable energy resources can be developed

What are the types of energy resources?

Renewable energy resources are ones that can be reused and therefore will not run out; they are known as infinite resources; examples include the wind and the Sun.

Non-renewable energy resources are resources that, once they have been used, can never be used again. As they took millions of years to form, they are known as finite resources; examples include coal and oil.

> **Renewable energy resources** can be used again and will not run out, such as the Sun or water.
>
> **Non-renewable energy resources** are finite in their supply. This means that there is a limit on the supply of the resource, such as coal.

The production and development of a non-renewable energy resource

Coal	Advantages	Disadvantages
Production	• Coal is found in many countries around the world. • Cheap to mine. Much of the coal which is mined around the world is just below the surface and can be mined very easily (for example, Australia).	• Waste heaps are left close to coal mines. • Deep-shaft mining can be dangerous, for example, nine miners died in China in May 2012 when a shaft collapsed.
Development	• It is relatively easy to convert it into energy by simply burning it. • Coal supplies should last for another 250 years.	• Acid rain is produced when coal is burnt. This has caused problems in the forests of Scandinavia. • When coal is burnt, greenhouse gases are emitted.

Figure 1 **Advantages and disadvantages of the production and development of coal as an energy resource**

The production and development of a renewable energy resource

Wind power	Advantages	Disadvantages
Production	• The wind is free. • Turbines are relatively cheap, costing £1,500 for a 1 kilowatt wind turbine.	• Some greenhouse gases are given off during the production of the turbine and when it is transported to its site.
Development	• New wind turbines are quiet and efficient. • It does not give off greenhouse gases. • Wind turbines can be on land or at sea.	• There needs to be an annual local wind speed of more than 6 metres per second. • They can be unsightly/visually intrusive. • Offshore turbines may disturb migration patterns of birds.

Figure 2 **Advantages and disadvantages of the production and development of wind power as an energy resource**

Countries use energy resources in different proportions (the energy mix)

The composition of the UK's energy mix

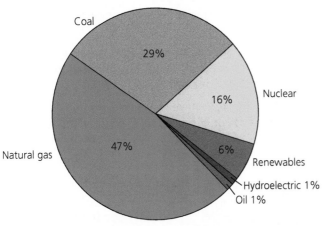

Figure 3 **Energy sources in UK, 2010**

> **Energy consumption** is how much energy is used.
>
> **Energy production** is how much energy is being made.
>
> **Global energy mix** refers to the way that countries use energy in different proportions.

	1961	1971	1981	1991	2001	2011
Coal	80	56	73	66	35	31
Gas	0	1	1	2	37	45
Hydroelectric	3	2	2	2	1	1
Nuclear	2	11	14	21	23	16
Oil sources	15	30	10	9	2	1
Renewables	0	0	0	0	2	6

Figure 4 **Changing UK energy mix, 1961 to 2011 (figures are percentages)**

Now test yourself

1 Draw a table which shows information on the production and development of coal and wind as energy resources. Give one advantage and one disadvantage for each resource.
2 Describe the changes in the use of coal and gas in the UK between 1961 and 2011.

TESTED

How are global variations in the energy mix dependent on a number of factors?

Factor	Effect
Population	Energy consumption has increased because: ● the world population increased by 27% between 1990 and 2008 ● the average use of energy per person globally increased by 10%. This is not equally spread around the world; countries with larger populations do not necessarily have the highest consumption of energy.
Wealth and income	Wealthy countries will use more energy than poor countries. If a country is wealthy, it will be able to provide the energy required by its population. The population will earn enough money to enable them to buy electrical equipment.
Availability of energy supplies	Countries that have a plentiful supply of energy may have high consumption. However, the rate of consumption also depends on other factors, such as the wealth and income of the population.

Figure 5 Factors affecting global variations in the energy mix

Country	Total energy consumption per capita per annum (2003) {kgoe/a}kg of oil equivalent	Total population in 2005 (millions)
Bangladesh	161	141
Brazil	1,067	186.4
Qatar	21,395	1
UK	3,918	60
Venezuela	2,057	26

Figure 6 Total energy consumption per capita and population of selected countries

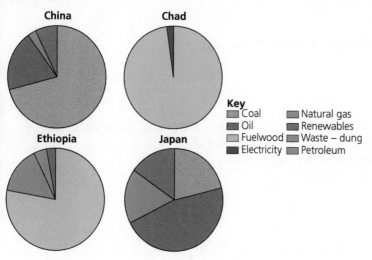

Figure 7 Energy mixes for China, Chad, Ethiopia and Japan

Key
- Coal
- Oil
- Fuelwood
- Electricity
- Natural gas
- Renewables
- Waste – dung
- Petroleum

Exam tip

You will need to be able to explain figures such as the ones in Figure 6.

Exam tip

You should be aware of the energy consumption of different countries in the world and be able to explain it.

Now test yourself

1 Qatar and Venezuela both have a plentiful supply of energy. Why is their consumption so different?
2 The UK has a smaller population than Bangladesh but its energy consumption is much higher. Why?
3 Explain why there are variations in the global energy mix between countries at different levels of development. Use information from Figure 7 in your answer.

TESTED

The increasing demand for energy is being met by renewable and non-renewable resources

How and why has the global demand for energy changed over the past 100 years due to human intervention?

REVISED

Population growth	In 1916 the world population was just under 2 billion, in 2011 it reached 7 billion. This growth in population has resulted in a great increase in the world's demand for energy. The majority of the population growth has been in developing, and more especially, emerging countries, causing a large increase in demand in these countries.
Increased wealth	The population of the world is becoming increasingly wealthy, which has enabled people to afford technology that uses energy. For example, 100 years ago very few people had cars; people heated their homes with coal and only heated part of the house. Nowadays central heating uses energy and many families in developed countries own two cars. In emerging countries, people's living conditions are improving as they become wealthier, which has meant that they are using more energy.
Technological advances	During the past 100 years there have been advances in technology which have required energy to power them.

Figure 8 **The impact of human intervention on the global demand for energy over the past 100 years**

Exam practice

Explain why energy consumption per person has increased over the past 100 years. (4 marks)

ONLINE

How and why has the global supply of energy changed over the past 100 years due to human intervention?

REVISED

Increased wealth and technological advances

- The increased wealth in the world has allowed the development of new energy sources and has therefore increased the supply of energy.
- It has paid for the development of new technology to search out new reserves of energy to provide for the increase in demand.
- New reserves of oil and gas were discovered under the sea bed in places such as the North Sea and off the coast of Venezuela, South America. New technology was developed to enable the reserves to be extracted.
- New technological advances have continued to open up new types of energy sources. Energy sources such as wind, solar energy and the power of water, both in rivers and the sea, have all been harnessed as new energy sources due to advances in technology.

Fracking is the process of drilling down into the earth to a gas-bearing rock. Water, sand and chemicals are blasted at the rock at high pressure which releases the gas inside the rock layers. The gas then flows out through the top of the well.

How are non-renewable energy resources being developed and what are the impacts on people and the environment?

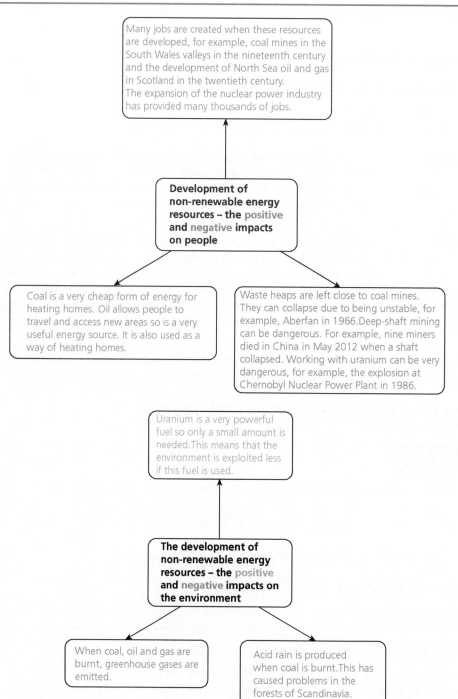

Many jobs are created when these resources are developed, for example, coal mines in the South Wales valleys in the nineteenth century and the development of North Sea oil and gas in Scotland in the twentieth century.
The expansion of the nuclear power industry has provided many thousands of jobs.

Development of non-renewable energy resources – the positive and negative impacts on people

Coal is a very cheap form of energy for heating homes. Oil allows people to travel and access new areas so is a very useful energy source. It is also used as a way of heating homes.

Waste heaps are left close to coal mines. They can collapse due to being unstable, for example, Aberfan in 1966.Deep-shaft mining can be dangerous. For example, nine miners died in China in May 2012 when a shaft collapsed. Working with uranium can be very dangerous, for example, the explosion at Chernobyl Nuclear Power Plant in 1986.

Uranium is a very powerful fuel so only a small amount is needed.This means that the environment is exploited less if this fuel is used.

The development of non-renewable energy resources – the positive and negative impacts on the environment

When coal, oil and gas are burnt, greenhouse gases are emitted.

Acid rain is produced when coal is burnt.This has caused problems in the forests of Scandinavia.

Exam practice

Assess the impacts on the environment of developing non-renewable and renewable energy resources.

(8 marks)

ONLINE

How are renewable energy resources being developed and what are the impacts on people and the environment?

REVISED

Fuel	Impacts on people (positive and negative)	Impacts on the environment (positive and negative)
Hydroelectric power	The reservoirs can be used for water sports or fishing. The reservoirs provide a water supply for areas that are located nearby. Often people have to move because their land is flooded to create the reservoir. The reservoirs are usually in remote areas, so people have to travel a long way to use the facilities.	It is less damaging to the environment as no greenhouse gases are produced. Large areas are flooded to create reservoirs to provide the water needed to drive the turbines. This will have an impact on the flora and fauna of the area.
Wind power	Home owners can have their own wind turbine so will benefit directly from 'free' energy. New wind turbines are quiet and efficient. They can be unsightly; many people see them as visually intrusive.	Wind turbines can be put out to sea so are less environmentally polluting. Wind turbines do not emit greenhouse gases once in use. They need a local wind speed of more than 6 m per second to be viable.
Solar power	They have no running costs so are a cheap source of energy. They can be fitted on homes so home owners can have their own energy supply. They can be unsightly; many people see them as visually intrusive.	Can be fitted on to roofs so do not take up extra land space. Solar panels do not emit greenhouse gases once in use. If put in fields they are taking up valuable space where food could be grown.

Figure 9 **The impacts of renewable energy resources on people and the environment**

How can technology (fracking) resolve energy resource shortages?

REVISED

At its present rate of consumption, the UK will run out of its own supplies of oil in 4.5 years' time. This makes the discovery of fracking an important resolution to the problem of energy security in the UK. However, there are a number of issues around fracking:

- It is a very water-intense process, which may bring it into conflict with other water users, such as agriculture.
- The fuel is also very high in carbon compared to renewables and will add to the greenhouse effect. It equals coal in CO_2 emissions.
- A lot of energy is needed to obtain the fuel; the energy input to energy output ratio is a very low net gain.

Exam tip

Fracking is a very controversial process. Make sure that you know its advantages and disadvantages.

Now test yourself

Define the term 'fracking'.

TESTED

Meeting the demands for energy resources can involve interventions by different interest groups

What are the attitudes of different stakeholders to the exploitation and consumption of energy resources?

REVISED

Stakeholders	Exploitation	Consumption
Government minister		'If people continue to use oil, we will raise a great deal of money in taxes to help develop better hospitals and schools.'
Chief executive from the oil industry	'There is plenty of oil left. We want to continue to extract oil because we provide lots of jobs.'	
Greenpeace campaigner		'We must build new homes with their own energy sources, such as solar panels, and plenty of insulation, so that each house consumes less energy.'
Land owner (farmer)	'If I put solar panels on my land, the government will pay me a grant. It is easier than farming and I make more money.'	
American Cadillac owner		'Technology has got us out of messes before. Soon something will be invented to solve the problem of dwindling oil supplies. I've been reading in the papers about something called 'fracking'.'
Saudi Arabian government	'Our country has developed using oil money. We must continue to exploit this resource if our country is to maintain its wealth.'	
Friends of the Earth energy campaigner	'Fracking for shale gas is not the solution to the UK energy crisis. We need energy based on renewables, not more fossil fuels which will add to greenhouse gas emissions.'	

Figure 10 Differing views on the exploitation and consumption of energy resources

Exam tip

You could be asked about people's opinions. Learn some of them in Figure 10.

Stakeholder is someone with an interest in the outcome of something.

Management and sustainable use of energy resources are required at a range of spatial scales from local to international

Why do renewable and non-renewable energy resources require sustainable management?

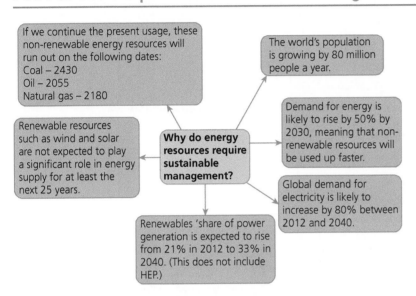

If we continue the present usage, these non-renewable energy resources will run out on the following dates:
Coal – 2430
Oil – 2055
Natural gas – 2180

The world's population is growing by 80 million people a year.

Renewable resources such as wind and solar are not expected to play a significant role in energy supply for at least the next 25 years.

Why do energy resources require sustainable management?

Demand for energy is likely to rise by 50% by 2030, meaning that non-renewable resources will be used up faster.

Global demand for electricity is likely to increase by 80% between 2012 and 2040.

Renewables 'share of power generation is expected to rise from 21% in 2012 to 33% in 2040. (This does not include HEP.)

> **Sustainable management** is using energy resources in a way which ensures that they are not exploited and hopefully will be able to meet the needs of future generations.

Exam practice

Explain **one** reason for the management of non-renewable energy resources.   (2 marks)

ONLINE

What are the different views held by individuals, organisations and governments on the management and sustainable use of energy resources?

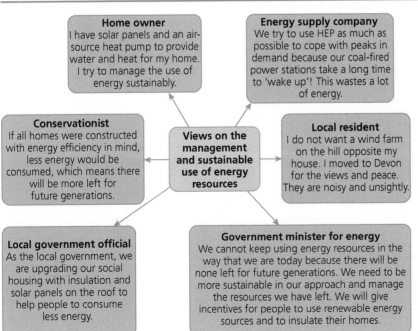

Home owner
I have solar panels and an air-source heat pump to provide water and heat for my home. I try to manage the use of energy sustainably.

Energy supply company
We try to use HEP as much as possible to cope with peaks in demand because our coal-fired power stations take a long time to 'wake up'! This wastes a lot of energy.

Conservationist
If all homes were constructed with energy efficiency in mind, less energy would be consumed, which means there will be more left for future generations.

Views on the management and sustainable use of energy resources

Local resident
I do not want a wind farm on the hill opposite my house. I moved to Devon for the views and peace. They are noisy and unsightly.

Local government official
As the local government, we are upgrading our social housing with insulation and solar panels on the roof to help people to consume less energy.

Government minister for energy
We cannot keep using energy resources in the way that we are today because there will be none left for future generations. We need to be more sustainable in our approach and manage the resources we have left. We will give incentives for people to use renewable energy sources and to insulate their homes.

How has Norway attempted to manage its energy resources in a sustainable way?

Oil and gas production	Renewable energy resources production
• Norway is a country in Northern Europe. It has vast reserves of oil and gas under the North Sea which it extracts and exports to other countries in Europe. • It is the world's sixth largest exporter of oil and second largest exporter of gas. • When Norway extracts oil and gas, only 54% is lost in the process; from most oil fields, up to 65% is lost. • In 2001 it produced 3.4 million barrels; in 2013 it produced approximately half of this amount to manage its resources sustainably.	• Norway is a mountainous country with reliable rainfall totals. • 99% of its electricity is produced from HEP. • Wind power is now being developed. Eight new wind farms are to be built onshore.

Figure 11 **How Norway has attempted to manage its energy resources in a sustainable way**

The management of Norway's energy consumption

Households
- No oil boilers.
- 40 to 60% of the house must be heated by means other than electricity.
- Education programme for children aged 9 to 12 years.
- Low-energy homes receive loans and grants which are available for energy efficiency measures.
- Heat pumps and biomass boilers get grant support.
- Helpline offering energy saving advice.

Industry
- Grants to install measures for energy recovery or use of waste heat in industrial processes.
- Energy information helpline.
- Grants to install renewables such as heat pumps.

Enova SF is a government organisation to promote energy savings. Norway set a target to reduce greenhouse gas emissions by 30% and to increase renewable energy share of total energy consumption to 67.5% by 2020. The incentives to reduce energy consumption include:

Transport
- Incentives to buy electric cars.
- Tolls on roads to stop people using them and speed limits.
- Car tax is higher on cars which are less fuel efficient.
- Cities which improved their public transport systems get grants from the government.
- Electrification of the country's rail network.

Enova SF is a Norwegian government enterprise responsible for the promotion of environmentally friendly production and consumption of energy.

Hydroelectric power (HEP) is energy produced by water turning a turbine to produce electricity.

Exam tip

Remember, just learn a few points, either numbers or places, for each of these case studies.

How has Bhutan attempted to manage its energy resources in a sustainable way?

- Bhutan is a country in South Asia.
- It has China to the north and India to the west, south and east.
- It has a population of approximately 750,000.
- Its rainfall varies between 500 and 5000 mm annually.
- The country has developed HEP as an energy source.
- The country is made up mainly of subsistence farmers and there is little industry.
- It has no reserves of oil or gas, so oil is imported.
- About 60 per cent of the population live in rural areas and use fuelwood as their main energy source.

Energy source	Usage
HEP	The Asian Development Bank has provided money to build dams and construct the grids to improve the electricity supply to rural areas in Bhutan. By 2013 Bhutan had provided electricity to 95% of its households.
Other micro-energy schemes	The other 5% of households live in remote rural areas. The government invested in small renewable energy schemes to bring electricity to these remote places to allow the whole of the country to benefit from electricity.
Fuelwood	The majority of the population, in both urban and rural areas, use fuelwood as the main domestic source of energy both for heating and to cook on. This is even with a supply of electricity, because fuelwood is freely available in the forests of Bhutan.

Figure 12 **How Bhutan has attempted to manage its energy resources in a sustainable way**

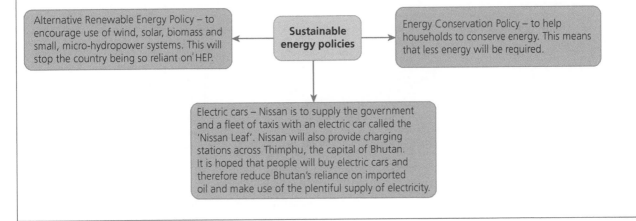

Exam tip

There is more information on Bhutan in your student textbook.

13 Water resource management

The supply of fresh water varies globally

How does the availability of fresh water vary on a global, national and local scale?

- The availability of fresh water is the amount of fresh water that can be easily accessed for human consumption, whether it is for domestic, industrial or agricultural usage.
- The amount of rainfall an area receives annually affects the availability.
- Another factor that restricts the amount of fresh water available is the fact that much of the water is away from the main concentrations of population.
- It has been estimated that the amount of fresh water available for human consumption is between 12,500 km^3 and 14,000 km^3 a year.

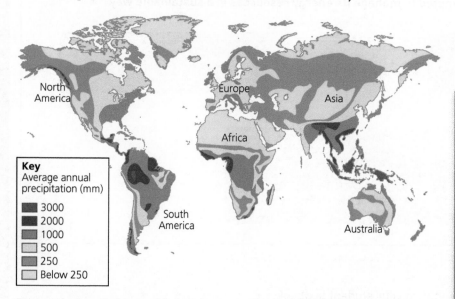

Key
Average annual
precipitation (mm)

- 3000
- 2000
- 1000
- 500
- 250
- Below 250

Figure 1 **Map of the world showing average annual precipitation**

Physical water scarcity is the term that applies to dry arid regions where fresh water naturally occurs in low quantities.

Fresh water is water that contains less than 1,000 milligrams per litre of dissolved solids, most often salt.

Economic water scarcity is the term that applies to areas that lack the capital or human power to invest in water sources and meet local demand; water is often available for people who can pay for it but not for the poor.

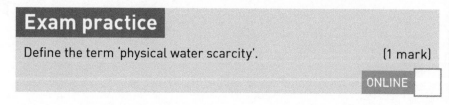

Exam practice

Define the term 'physical water scarcity'. (1 mark)

ONLINE

Why do some parts of the world have a water surplus or a water deficit?

- A number of factors affect whether a region has too much or too little water.
- A region can have a physical water surplus or water deficit; this relates to the amount of rainfall it receives.
- Another factor is the evapotranspiration rates. Some regions with a reasonable amount of rainfall have very high evapotranspiration rates. The water does not have time to enter water sources that make it available for human consumption.
- A region can have an economic surplus or deficit; this relates to whether the government of an area can afford to provide water supply to the population.

Domestic usage is the use of water by households.

Industrial usage is the use of water by factories or the companies who produce energy. It is also the water used in offices and schools.

Agricultural usage is the use of water by farmers to water their crops or animals.

Water surplus is when water supply exceeds water demand.

Water deficit is when water demand exceeds water supply.

Evapotranspiration is the movement of water from the land to the atmosphere by evaporation and plant transpiration.

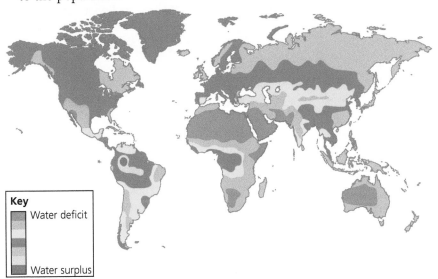

Key
- Water deficit
- Water surplus

Figure 2 Map of the world showing areas of water surplus and deficit

Now test yourself

Research and learn the names of countries which have:
- high rainfall
- low rainfall
- water surplus
- water deficit.

These may be the same countries or they may be different ones.

Exam tip

Exam questions are unlikely to expect you to remember areas of the world, but you should know the continents and the names of countries that have extremes of rainfall and water supply.

How and why has the supply and demand for water changed in the past 50 years due to human intervention?

Water supply

Emerging or developing countries	Developed countries
The supply of piped fresh drinking water to households has increased in the past 50 years. This has been carried out by charitable organisations, such as Water Aid UK and World Health Organisation, which have enabled people to gain access to improved drinking water. In 2015, 8% of the global population still did not have access to clean drinking water.	The supply of fresh water to households in developed countries has changed little over the past 50 years, although the variations in rainfall totals have had an impact on the amount of fresh water available in some developed countries.

Figure 3 **Changes in the supply of water in the past 50 years**

Water demand

There has been an increase in the demand for water globally for a number of reasons:

- an increase in manufacturing industry in developing and emerging countries
- an increase in thermal electricity generation
- an increase in domestic use
- an increase in meat production; meat production uses eight to ten times more water than cereal production

- an increase in water for irrigation
- it takes an average 3,000 litres of water to produce one person's daily food intake and the population of the world continues to grow.

Emerging or developing countries	Developed countries
As the supply of piped fresh water has improved to households in these countries, then the demand for water has also increased. This means that the country is using its fresh water which in the past it did not have access to. This could possibly create water shortages. For example, in China fresh water has been supplied to many households over the past 50 years. This has caused an increase in demand in areas of the country which have less rainfall because this is where the population lives. The Chinese have started to build large water transfer schemes to deal with the surplus of water in some areas of the country and the deficit of water in others.	As developed countries have become wealthier, the demand for water has increased. This is due a number of factors: - Technological advances: dishwashers and washing machines use much more water than washing dishes and clothes by hand. The average dishwasher uses 3,000 litres of water a year. - Changes to personal hygiene: 50 years ago houses did not all have bathrooms; many houses now have more than one bathroom. - Sport has increased: for example, golf, which uses large amounts of water to keep the course green. - Leisure has increased: the building of swimming pools in people's gardens has increased.

Figure 4 **Changes in the demand for water in the past 50 years**

Exam practice

Explain why the demand for water has increased over the past 50 years.

(4 marks)

ONLINE

Developing and developed countries have different water consumption patterns

What is the proportion of water used by agriculture, industry and domestic in developed countries and emerging or developing countries?

REVISED

Use	World %	Developed %	Developing %
Agriculture	70	30	82
Domestic	8	11	8
Industry	22	59	10

Figure 5 **World water usage**

Irrigation is the artificial watering of the land.

Cottage industry is small-scale production, often in a room of a person's home.

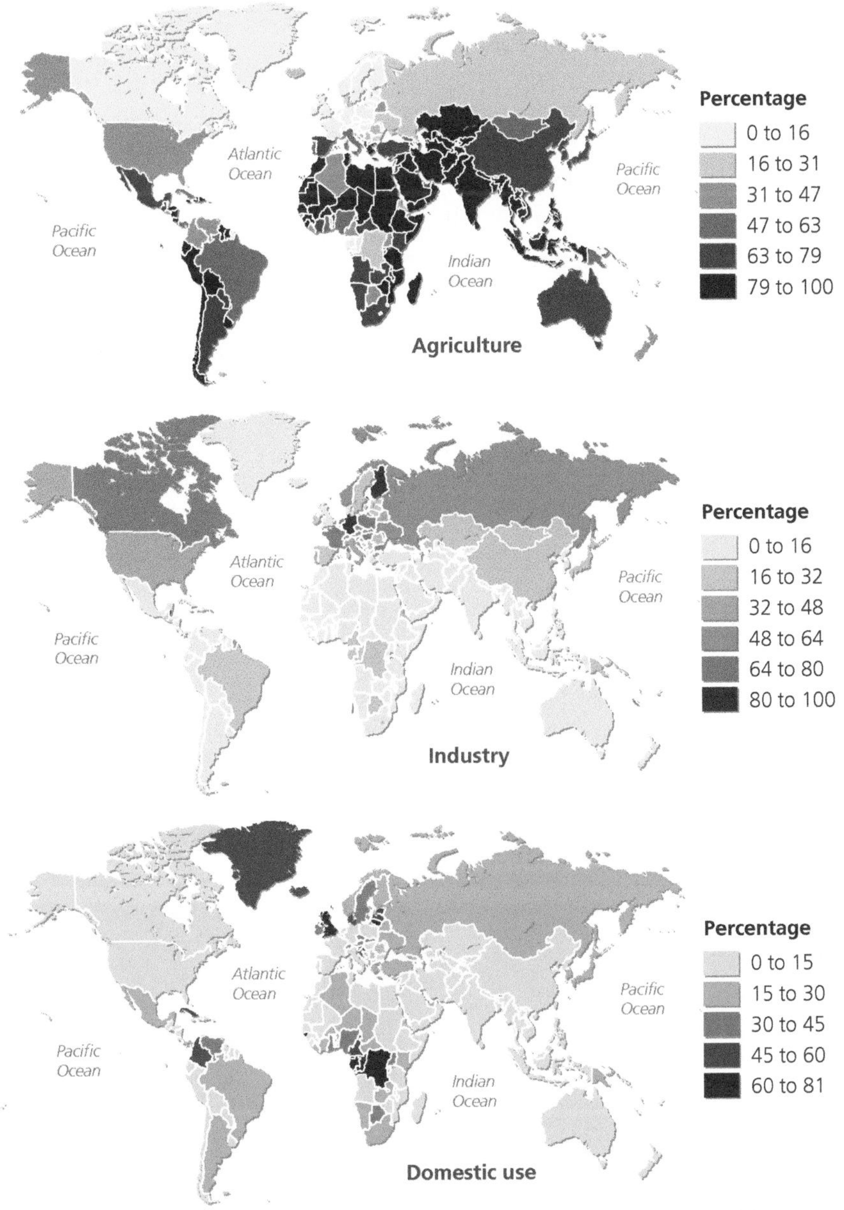
Atlantic Ocean
Pacific Ocean
Pacific Ocean
Indian Ocean
Agriculture
Percentage
0 to 16
16 to 31
31 to 47
47 to 63
63 to 79
79 to 100
Atlantic Ocean
Pacific Ocean
Pacific Ocean
Indian Ocean
Industry
Percentage
0 to 16
16 to 32
32 to 48
48 to 64
64 to 80
80 to 100
Atlantic Ocean
Pacific Ocean
Pacific Ocean
Indian Ocean
Domestic use
Percentage
0 to 15
15 to 30
30 to 45
45 to 60
60 to 81

Figure 6 **Agriculture, domestic and industry water usage in 2000**

Exam tip

You should be aware of the water consumption of different countries in the world and be able to explain it.

Why are there differences in water usage between developed and emerging or developing countries?

REVISED

	Developed	Emerging and developing
Domestic	People have water piped into their homes. This allows them to use water in washing machines and dishwashers. The 'showering society' means many people shower every day.	Many people do not have piped water to their homes so they wash their clothes and dishes in local streams. The same water is used for many different purposes.
Agricultural	Irrigation systems are used which are operated by computers. The computer can determine exactly how much water is required and supply the water quickly, up to 75 litres per second.	Plants are watered using buckets or very simple irrigation systems, which supply water slowly at approximately 1 litre per second.
Industrial	Factories are on a large scale and use thousands of litres of water.	Small-scale cottage industries use only a small amount of water.

Figure 7 Differences in water usage between developed and emerging or developing countries

Now test yourself

TESTED

Explain why there are differences in water usage between developed and developing countries.

Countries at different levels of development have water supply problems

Why does the UK have water supply problems?

REVISED ☐

Problem	Description of problem
Imbalance of the supply from rainfall and the demand from population	The rainfall that is received by the UK is very varied. The north and west of the country receive the highest amounts, with Keswick in the Lake District receiving on average 1,500 mm of rain a year and London in the Southeast only receiving 550 mm a year. This means that the supply is plentiful in the north and west. However, one-third of the population of the UK lives in the Southeast which is the driest part of the UK.
Ageing infrastructure: leakage to sewage and water pipes	Many of the water pipes which supply water to households and industry in the UK are over 100 years old. The ageing pipes do not cause a problem with the quality of the water but there is a problem with leaks. In 2009, water firms in the UK lost 3.29 billion litres of water because of leaks; this is a third of the water that was supplied. Between 2004 and 2009, Thames Water reduced leaks by replacing old pipes by 27% at a cost of £1 billion. The sewerage system of the UK also has an ageing infrastructure which, in many places, is over 100 years old. Before October 2011, much of the sewerage network was owned by the people who lived in the street, possibly without them knowing this! On 1 October 2011, most of the ownership of the sewerage network passed to the water companies, making them responsible for leaks.
Seasonal imbalances	The UK receives most of its rainfall in the winter but the highest water demands are in the summer. This can cause a problem of water supply for the water companies especially if the country has experienced a dry winter or spring.

Figure 8 **The UK's water supply problems**

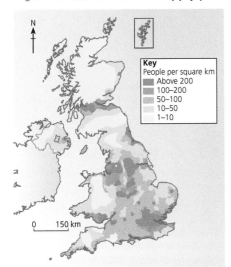

Figure 9 **The population density of the UK**

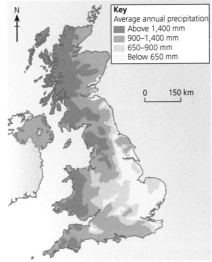

Figure 10 **UK average rainfall**

Exam practice

Explain **one** water supply problem in the UK.

(2 marks)

ONLINE ☐

Why do emerging and developing countries have water supply problems?

Problem	Description of the problem
Lack of clean piped water	Approximately 1 billion people in developing and emerging countries do not have safe water to drink.
Water-borne disease	Diseases such as dysentery are caused by drinking water that is not clean. Young children and older people are particularly susceptible to diarrhoeal diseases, which are related to drinking dirty water, but these deaths, especially among children, decreased by nearly a million between 1990 and 2012.
Pollution	Many people in emerging and developing countries still use rivers for their drinking water. In the Amazon region, the indigenous tribes' river water has been polluted due to mining and oil extraction. The waste materials from these industries are washed into the rivers. The pollution can cause an increased risk of cancer, miscarriage, headaches and nausea. This is because their drinking water now contains toxins way above the level acceptable for human consumption.
Low annual rainfall	Many developing and emerging countries are in parts of the world that have a low annual rainfall. This means that, as the population increases, these countries will have a physical scarcity of water. It is estimated that by 2025, 1.8 billion people will be living in countries with water scarcity.

Figure 11 **Water supply problems of emerging and developing countries**

Meeting the demands for water resources could involve technology and interventions by different interest groups

How do attitudes to the exploitation and consumption of water resources vary with different stakeholders?

REVISED

Stakeholders	Attitude
Head of Coca-Cola, India	'If we are to keep our plants working efficiently and provide jobs for the local people we need to use a lot of water. If this has to come from ground water sources this is not our fault; we have a business to run.'
Government of India	'The Coca-Cola plants in our country will close if they are in areas of water shortage. We need to look after the welfare of our people and conserve water for domestic use.'
Water Aid campaigner	'Many countries in the world are short of water. We need to conserve water and use it sensibly if there is to be enough to go round.'
Farmer in Arizona, USA	'We need to irrigate our crops so that we can make a profit and, of course, provide enough food for the people of America.'
HEP producer	'I need to dam the river if I am to create enough energy. It cannot be helped if this stops the river flooding, which used to provide irrigation water for the farmers downstream.'
Major of London	'If we keep using groundwater supplies at the present rate, there will soon be none left. We have to start conserving our water supplies or there will be severe water shortages in dry years.'
London resident	'They keep going on about water shortages. I cross the Thames every day on the way to work. There seems to be plenty of water in it. What about all the floods that happened last year!'

Figure 12 **Differing views on the exploitation and consumption of water resources**

Now test yourself

TESTED

Define the term 'desalination'.

Exam tip

You could be asked about people's opinions. Learn some of them in Figure 12.

How can technology resolve water resource shortages?

REVISED

- There are currently 16,000 desalination plants worldwide, producing roughly 70 million cubic metres of fresh drinking water per day.
- Saudi Arabia is the country which has the most desalination plants, with the USA in second place.
- The biggest problem with desalination is that it takes a lot of energy to desalinate a litre of sea water. However, the plants get around this by either using their country's cheap supplies of oil and gas, as is the case in the Gulf states, by using cheaper night-time electricity or, more recently, by using solar power to operate the plants.

Exam tip

Desalination is a controversial process. Make sure that you know its advantages and disadvantages.

Desalination is the removal of minerals from salt water to make it drinkable.

Stakeholder is someone with an interest in the outcome of something.

The plant can produce 150 million litres of water a day.

Thames Water built a desalination plant at Beckton on the Thames Estuary following droughts in the 1990s.

It takes water from the Thames.

The plant started to produce fresh water in 2010.

It is not used everyday but is used if there is a drought or levels in reservoirs or ground water supplies are low.

It is powered 100% by renewable energy.

The total cost of the scheme was £250 million.

Figure 13 **The Beckton desalination plant**

It produces just over 1 million m³ of fresh drinking water a day.

The population of Saudi Arabia has quadrupled in 40 years and the government needs to supply its population with fresh water.

It takes water from the Persian Gulf.

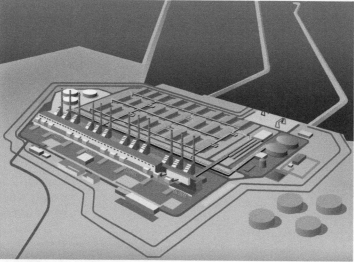

It cost £3.5 million to build.

It started to produce fresh water in 2014.

It takes 2,400 megawatts of electricity a day to power the plant.

It is powered by solar energy during the day, which is when the peak supply is needed.

Figure 14 **Saudi Arabian desalination plant on the Persian Gulf**

Exam tip

Choose either the Beckton desalination plant or the Saudi Arabian desalination plant. Learn the information shown in your chosen plant.

Exam practice

Explain the disadvantages of desalination. (3 marks)

ONLINE

Management and sustainable use of water resources are required at a range of spatial scales from local to international

Finite resource is something that there is a definitive amount of.

Sustainable water management is meeting consumer demand for water now without compromising the needs of future generations.

Aquifers are water-bearing rocks.

Metered households are households in which meters measure the amount of water that the household uses and the homeowner pays for each drop of water.

Unmetered households are households which do not have a meter; the water company estimates how much water the householder will use in a year and charges for this amount.

Why do water resources require sustainable management?

REVISED

2.1 billion people were given access to clean drinking water between 1990 and 2011. 800 million are still without access to clean water.

The world's population is growing by 80 million people a year. This is an increase of 64 billion cubic metres of water a year.

Every day, 2 million tonnes of human waste are disposed of in water courses.

Why do water resources require sustainable management?

There is an increase in demand for meat globally. Meat production requires 8 to 10 times more water to produce than cereal crops.

In developing and emerging countries, 70% of industrial waste is dumped untreated into rivers.

By 2050 the demand for water globally is likely to increase by 55%.

70% of water that is withdrawn from rivers globally is by agriculture for food production. Irrigated land represents 20% of the land used for farming but contributes 40% of the global food crop.

Now test yourself

State **three** reasons why water resources require sustainable management.

TESTED

What are the different views held by individuals, organisations and governments on the management and sustainable use of water resources?

REVISED

Government of China
We need to provide fresh water for our people but will there be enough to go round?

Resident of a developing country
My life has been so much better since the government provided me with piped water.

UK resident
Why shouldn't I use as much water as I like? It never stops raining!

Conservationist
We need to work together to integrate water management so that we can get the most out of each river catchment area.

Views held by individuals, organisations and governments on the management and sustainable use of water resources

Chinese farmer
I need to use water to irrigate my crops or the harvest will fail and I will make no money. I am pleased that I have been shown the new techniques to irrigate my fields which take less time and use less water.

Thames Water Authority
We are working as hard as we can to replace old Victorian water mains and to mend leaks as soon as they are reported to us.

How has the UK attempted to manage its water resources in a sustainable way?

	Unmetered households (litres per day)	Metered households (litres per day)	All households (litres per day)
2000	152	134	149
2002	157	137	150
2004	154	136	149
2006	153	134	147
2008	152	130	145
2010	158	129	144

Figure 15 **Domestic water usage**

- The UK has a plentiful supply of water due to the amount of rainfall it receives.
- Areas with high rainfall have low population densities.
- For many years, water has been transferred around the UK from areas which receive plenty of rainfall to areas with high populations. For example, the Elan Valley has supplied water for Birmingham since the early 1900s.
- Further and larger water transfer schemes would cost a lot to build and pumping the water around the country would be very expensive.
- Water is also obtained from aquifers in the ground, which are water-bearing rocks. Over the last 30 years this supply has not been replenished as fast as it is being used by water companies.
- There is also the uncertainty of climate change which could bring more rainfall or it could mean more droughts. As nothing is certain, decisions have been taken to encourage changes to the way that we view water.

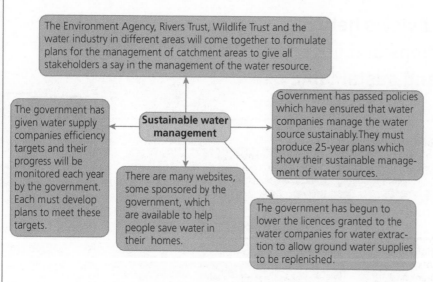

Exam tip

Remember, just learn a few points, either numbers or places, for each of these case studies.

How has China attempted to manage its water resources in a sustainable way?

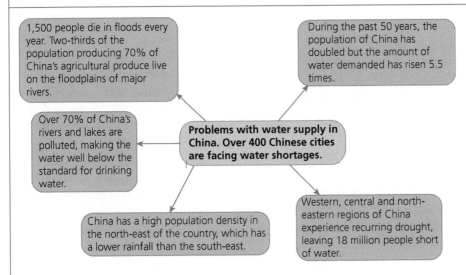

1,500 people die in floods every year. Two-thirds of the population producing 70% of China's agricultural produce live on the floodplains of major rivers.

During the past 50 years, the population of China has doubled but the amount of water demanded has risen 5.5 times.

Over 70% of China's rivers and lakes are polluted, making the water well below the standard for drinking water.

Problems with water supply in China. Over 400 Chinese cities are facing water shortages.

China has a high population density in the north-east of the country, which has a lower rainfall than the south-east.

Western, central and north-eastern regions of China experience recurring drought, leaving 18 million people short of water.

Solutions to the water shortage problem

The government of China has initiated a number of policies to manage the available water resources sustainably to ensure that there is enough for their population and for future generations.

Desalination plants are being built on China's coastline to provide water for cities in the north and east. £2.1 billion is being spent on desalination plants, which will triple the amount of water available for human use by 2020.

- Water is being redirected from China's wetter south to the north of the country which has a physical scarcity of water.
- Projects are being carried out in rural China by charities to cut the amount of water used to produce food. They are working with farmers to improve irrigation methods.
- The government has introduced a water-saving campaign using Olympic athletes to go into schools to teach the children the importance of saving water.
- Beijing has a water conservancy museum which aims to show people how much water they use in their daily life by using Coca-Cola bottles.
- In Shanghai 50 wells have been dug 240 m deep beneath large residential areas and universities. The water from these wells will be used when there are acute water shortages in the city.
- China has also spent money on improving its reservoirs. By the end of 2015, over 50,000 reservoirs had been reinforced and their water quality improved.

Exam tip

The UK and China are the two case studies in this chapter. You will need to learn how they have attempted to manage their water resources in a sustainable way. One way is to produce sticky notes with the information on. Don't add too much detail, just the facts that you are going to learn.

14 Geographical investigations: fieldwork

As part of your course you have completed two geographical enquiries – one piece of physical geography fieldwork and one piece of human geography fieldwork. You will have written up both pieces of work as an enquiry. But how do you learn the information so that you can answer questions on an examination paper?

One way is to break the information down into the route to enquiry and prepare answers for the sorts of questions you can be asked using your own fieldwork experiences. The examiner will be looking for information about *your* study, not general comments about any study. Therefore, you must revise your fieldwork as you do the content for Papers 1 and 2.

> **Exam tip**
>
> Learn the key terms in the box. You could put them on sticky notes around your mirror to help you to remember them.

Exam practice

What is the difference between qualitative and quantitative techniques?

(2 marks)

ONLINE

> **Exam tip**
>
> Perhaps obvious, but be sure that you know which is your physical and which is your human investigation!

Fieldwork questions

REVISED

You will need to know the questions you developed. You may be asked to write out your questions and then justify your choice. You would do this by referring to the area where you completed the study and why the question was appropriate for that particular area.

For example, for your physical investigation:

Does the discharge of the River X increase with distance from its source to its confluence with the River Y?

or

Does the beach increase in gradient as you move from the shoreline to the sand dunes?

Now test yourself

TESTED

1 What questions did you devise for your physical and human investigations?
2 Justify your choice of questions.

Quantitative techniques are data collection techniques which record statistical data and/or measurements and are carried out in the field.

Qualitative techniques are techniques where information is gained through observation. They usually involve a description of a feature.

Primary data source is a piece of evidence which is first hand, it is collected by the researcher themselves.

Secondary data source is evidence which is collected by someone else.

Discharge is the amount of water passing a specific point at a given time and is measured in cubic metres per second. It is calculated by cross-section area x velocity.

Random sampling is where each member of the population has an equal chance of being chosen.

Stratified sampling is when the sample contains an equal number of results in each category. For example, if a questionnaire was being carried out in a tourist resort and the views of both residents and tourists were required. There would have to be an equal number of each in the results.

Systematic sampling is when the sample is collected according to an agreed sample. For example, every fifth person might be interviewed or every tenth house might be used in the street sample.

Fieldwork techniques and methods

There are certain data collection techniques that you have been asked to concentrate on for each of the physical and human investigations. Make sure that you know everything about these techniques.

Now test yourself

Look at the table below. For the two investigations that you are revising, make sure you can answer the following questions about the techniques you used:

1 How did you carry out the technique?
   a) How did you use the equipment?
   b) If you were sampling, what was your method?
2 Why did you use this technique? (justification)
3 How successful was this technique? (evaluation)

Investigation	Fieldwork technique	Examples of techniques and methods
Rivers	**Quantitative technique** –measuring river discharge	Use a flow meter to measure water velocity. Measure a cross-section area of the river channel (width and depth). Calculate the river discharge: area x velocity = discharge
	Qualitative technique – recording landforms that make up the river landscape	Field sketches – include annotations Photographs – add descriptions
Coasts	**Quantitative technique** – measuring beach morphology and sediment characteristics	Measure beach profiles using ranging roles and a clinometer. Measure sediment characteristics – the size and shape of pebbles – using the Powers roundness index to measure samples along a beach.
	Qualitative technique – recording landforms that make up the coastal landscape	Field sketches – include annotations Photographs – add descriptions
Urban	**Quantitative technique** – measuring land use function	Carry out a land-use survey
	Qualitative technique – recording the quality of the urban environment	Field sketches – include annotations Photographs – add descriptions Environmental quality survey
Rural	**Quantitative technique** – measuring flows of people within a rural settlement	Counting pedestrians/traffic at certain points in the settlement for fifteen minutes every hour. This data can be collated and flow maps can be produced
	Qualitative technique – recording the views of people on the quality of the rural environment	Field sketches – include annotations Photographs – add descriptions Environmental quality survey

Figure 1 Fieldwork techniques and methods

Exam practice

1 State the sampling method used in your physical environment investigation. (1 mark)
2 Explain one reason why you used this technique. (2 marks)

Human interaction with the physical environment

REVISED ▢

You could be asked about the implication of the physical environment for the people living in the area on either of the investigations you have undertaken. In all cases, the examiner will be looking for evidence of your own study. You will not be credited for just naming the area. You must ensure that the examiner has evidence that you actually went on the study and know the area.

Now test yourself

For your investigations, answer the question in the table below. Remember to name particular places and use your results.

TESTED ▢

Investigation	Focus	Question
Rivers	The implications of river processes for people living in the catchment area	Is there a risk of flooding from the river you have studied?
Coasts	The implications of coastal processes for people living in the catchment area	Is there a problem with longshore drift removing beach materials?
Urban	The interaction between physical landscape features, the central/inner urban area and residents and visitors	Have physical features influenced the use of land in the central urban area?
Rural	The interaction between physical landscape features, rural settlements and residents and visitors	Has the physical environment influenced the flows of people in the area?

Figure 2 **Investigation questions and responses**

Analysis and conclusions

REVISED ▢

In the examination, you could be given the results from a student's notebook and asked to analyse and draw conclusions from those results. Another type of question could ask about your conclusions. However, it is unlikely to be a straightforward question. It is more likely to ask you to evaluate your conclusions. Therefore, you must know your conclusions and why you came to those conclusions, using evidence from your study which you have revised.

Exam practice

Making reference to your results, suggest how your investigations could be improved. (2 marks)

ONLINE ▢

Secondary data

REVISED ▢

There are specific secondary data sources that you will need to know about for each of the physical and human investigations, along with one other secondary source.

Investigation	Secondary data source	Examples
Rivers	A flood risk map	Environment Agency flood risk maps can be used to assess the impact of the river on people who live in the area.
	One other secondary source	Newspaper articles with information on floods in the area.
Coasts	A geology map	The British Geology Survey's Geology of Britain viewer can be used to investigate rock types of the study area.
	One other secondary source	Newspaper articles could be used to find out information on coastal recession in the area.
Urban and Rural	Census data	Office for National Statistics (ONS) Neighbourhood Statistics can be used to get information on people who live in the study area. This information may be used to investigate why changes have occurred to the local landscape.
	One other secondary source	Newspaper articles could be used to find out information about what the area was like in the past. GOAD maps of the area could be used to find out about previous land uses.

Figure 3 **Secondary data**

Now test yourself and exam practice answers at **www.hoddereducation.co.uk/myrevisionnotes**

15 Geographical investigations: UK challenges

In Section C of Paper 3, you will be required to draw on the knowledge and understanding you have gained in the whole course to investigate a contemporary challenge facing the UK. The challenge will be on one or more of the following four themes.

Theme	Content	Related content	Key words
1 The UK's resource consumption and environmental sustainability challenge	Changes in the UK's population in the next 50 years and implications on resource consumption.	Pages 42, 54	**Ecosystem** is a community of plants and animals and their non-living environment.
	Pressures of growing populations on the UK's ecosystems.	Pages 54, 55, 72, 96	**Resource** is a stock or supply of something that is useful to people.
	Range of national sustainable transport options for the UK.	Page 75	
2 The UK settlement, population and economic challenges	The 'two-speed economy' and options for bridging the gap between south east and the rest of the UK.	Page 67	**Greenfield development** is when houses and industry are built on land at the edge of the city which has never been built on before.
	Costs and benefits of greenfield development and the regeneration of brownfield sites.	Pages 71, 74	**Brownfield site** is land within a city which is no longer used; it may contain old factories or housing, or it may have been cleared for redevelopment.
	UK net migration statistics and their reliability and values and attitudes of different stakeholders towards migration.	Page 72	**Stakeholders** towards migration are people who are either involved in emigration or who will be affected by migration.
			Disposable income is the money people have left to spend after essential goods and services have been paid for.
			Two-speed economy refers to the fact that the Southeast has a faster rate of growth than the rest of the UK.
3 The UK's landscape challenges	Approaches to conservation and development of UK National Parks	Pages 34, 36	**National park** is an area of countryside which is protected because of its natural beauty and also managed for visitor recreation.
	Approaches to managing river and coastal UK flood risk.	Pages 17, 18, 19	**Conservation** is keeping something as it is and not changing it in any way to preserve it for future generations.
4 The UK's climate change challenges	Uncertainties about how global climate change will impact on the UK's future climate.	Page 43	**Climate change** is a long-term movement in the weather patterns and average temperatures experienced by the Earth.
	Impacts of climate change on people and landscapes in UK	Page 58	
	Range of responses to climate change in the UK at a local and national scale.	Page 127	

Figure 4 **Where to find information in this book**

Responses to climate change in the UK at a local and national scale

REVISED

- **By the government:** Many initiatives to encourage home owners and businesses to be more energy efficient and to use renewable fuels rather than traditional fossil fuels, for example:
  - the renewable heat incentive – grants for home owners using air or ground source heat pumps or thermal hot water heating on the roof of their homes
  - the 'feed in tariff' – extra electricity produced by solar panels on roofs feeds into the national grid and the home owner or business is paid for it
  - electric cars are exempt from road tax
  - grants available to enable home owners to insulate their homes for free
- **By schools:**
  - introducing energy-efficient water and central heating systems run from renewable sources such as wind turbines or solar panels
  - notices asking people to switch off lights
  - energy prefects who switch off lights and computers at the end of the day
- **By local councils:** councils introducing schemes to cut carbon emissions:
  - Giving away free low-energy light bulbs
  - Combined heat and power (CHP) schemes
- **By local interest groups:** For example, 'Manchester is my planet' runs a pledge campaign to encourage individuals to reduce their carbon footprint.

Exam practice

1 Discuss the implications for the use of resources of the changes to the UK's population during the next 50 years. [16]
2 Evaluate the different sustainable transport options which are used in the UK. [16]
3 Discuss the values and attitudes of different people towards migration and the reliability of UK net migration statistics. [16]
4 Evaluate the costs and benefits of greenfield development and building on brownfield sites. [16]
5 Assess the approaches which have been used to manage rivers and coastal areas of the UK. [16]
6 Discuss how global climate change will impact on the UK's future climate. [16]
7 Assess the impacts of climate change on people and landscapes in the UK. [16]

ONLINE